The Neuroanatomy Riddle Book

Series Authors
Dr James Dolbow
University Hospitals Cleveland Medical Center, Cleveland, Ohio, USA

Dr Neel Fotedar
University Hospitals Cleveland Medical Center, Cleveland, Ohio, USA

Dr Joshua Edmondson
University Hospitals Cleveland Medical Center, Cleveland, Ohio, USA

Also in this series:
The Neurology Riddle Book: 150 Fun and Challenging
Neurology Riddles
James Dolbow, Neel Fotedar
ISBN 9781009527361

The Neuroanatomy Riddle Book

150 Fun and Challenging Neuroanatomy Riddles

James Dolbow

University Hospitals Cleveland Medical Center

Joshua Edmondson

University Hospitals Cleveland Medical Center

Neel Fotedar

University Hospitals Cleveland Medical Center

CAMBRIDGE
UNIVERSITY PRESS

Shaftesbury Road, Cambridge CB2 8EA, United Kingdom

One Liberty Plaza, 20th Floor, New York, NY 10006, USA

477 Williamstown Road, Port Melbourne, VIC 3207, Australia

314–321, 3rd Floor, Plot 3, Splendor Forum, Jasola District Centre, New Delhi – 110025, India

103 Penang Road, #05–06/07, Visioncrest Commercial, Singapore 238467

Cambridge University Press is part of Cambridge University Press & Assessment, a department of the University of Cambridge.

We share the University's mission to contribute to society through the pursuit of education, learning and research at the highest international levels of excellence.

www.cambridge.org
Information on this title: www.cambridge.org/9781009527415

DOI: 10.1017/9781009527422

When citing this work, please include a reference to the DOI 10.1017/9781009527422

First published 2025

A catalogue record for this publication is available from the British Library

Library of Congress Cataloging-in-Publication Data
Names: Dolbow, James, author. | Fotedar, Neel, author. | Edmondson, Joshua, author.
Title: The neuroanatomy riddle book : 150 fun and challenging neuroanatomy riddles / James Dolbow, University Hospital Cleveland Medical Center, Neel Fotedar, University Hospital Cleveland Medical Center, Joshua Edmondson, University Hospital Cleveland Medical Center.
Description: Cambridge, United Kingdom ; New York, NY : Cambridge University Press, 2024. | Includes index.
Identifiers: LCCN 2024010728 (print) | LCCN 2024010729 (ebook) | ISBN 9781009527415 (paperback) | ISBN 9781009527422 (epub)
Subjects: LCSH: Neuroanatomy–Miscellanea. | Neuroanatomy–Humor. | Riddles.
Classification: LCC QM451 .D65 2024 (print) | LCC QM451 (ebook) | DDC 611/.8–dc23/eng/20240417
LC record available at https://lccn.loc.gov/2024010728
LC ebook record available at https://lccn.loc.gov/2024010729

ISBN 978-1-009-52741-5 Paperback

•••

Every effort has been made in preparing this book to provide accurate and up-to-date information which is in accord with accepted standards and practice at the time of publication. Although case histories are drawn from actual cases, every effort has been made to disguise the identities of the individuals involved. Nevertheless, the authors, editors and publishers can make no warranties that the information contained herein is totally free from error, not least because clinical standards are constantly changing through research and regulation. The authors, editors and publishers therefore disclaim all liability for direct or consequential damages resulting from the use of material contained in this book. Readers are strongly advised to pay careful attention to information provided by the manufacturer of any drugs or equipment that they plan to use.

This book is dedicated to our mentors.

Contents

Preface

So if you are anything like my colleagues and me, you love a puzzle, and more than that, you love a neurological puzzle. Well, the goal of this book is to provide these. In this book, you will find 150 four-line riddles with the same general structure and flow. Each of them describes a specific neuroanatomical structure in cryptic form.

This can be any of the following:

1. A lobe or general region of the brain
2. A central/cranial or peripheral nerve
3. A central or peripheral nucleus
4. A blood vessel supplying key neurological structures
5. A specific anatomical brain structure
6. A central nervous system tract
7. A neuroanatomical space or passage

This cannot be any of the following:

1. A non-neurological structure
2. A disease or pathology
3. A reflex, clinical finding, sign, or symptom.

All riddles will incorporate in some way a few of the following: hints to its general location in the body, what it branches from/to, its general function or dysfunction if impaired, a well-known clinical correlate, or its Latin or Greek name origin. If after reading each riddle, you are still stumped, at the bottom of that page you will find two hints to help guide you. On the following page, you will find the answer as well as a brief description of the structure, why that answer was correct, and some interesting anatomical/ clinical correlates.

Enjoy!

1

• • • • •

The smallest but the longest,
From the back I come to play,
Though my goal may be oblique,
Without me you'll tilt away.

Hint #1:

Think head tilt.

Hint #2:

Cranial nerve.

Trochlear Nerve (CN IV)

Cranial nerve IV, the trochlear nerve, is the *smallest* cranial nerve and has the *longest* intracranial path as it exits from the dorsal midbrain (at the level of inferior colliculus), crosses to the other side of the brainstem and travels along the lateral wall of the cavernous sinus, exiting the skull through the superior orbital fissure, and traveling all the way to the superior *oblique* muscle. Because of decussation of the intraparenchymal fibers after originating from the nucleus, the left trochlear nucleus eventually innervates the right superior oblique muscle and vice versa.

The superior oblique muscle acts as an intorter and depressor. Thus, a lesion of the nerve would inhibit this movement, creating a superiorly deviated orbit that is extorted, thus leading to a vertical and torsional diplopia. To correct for the misalignment, patients often *tilt* their head forward (to correct for elevation) and away from the affected eye (to correct for extortion).

A skew deviation can also produce a vertical misalignment of eyes like trochlear nerve palsy. A three-step Parks–Bielschowsky test can help localize the affected muscle in a patient with vertical diplopia. It can also differentiate between skew deviation and trochlear nerve palsy because the degree of vertical misalignment changes with gaze and head tilt in trochlear nerve palsy but not in skew deviation. Another way to distinguish between the two is by the upright-supine test. In a patient with skew deviation, there is >50% reduction in both vertical and torsional misalignments from the upright position to the supine position.

2

• • • • •

Two peas in a pod that lie side by side,
Near Greek "across" or "between,"
With pints come lies and paralyzed eyes,
And sideways you will lean.

Hint #1:

Diencephalon, which comes from the Greek *dia*, meaning "between/across," and *enkephalos*, meaning "brain."

Hint #2:

Alcohol is commonly served in a pint glass.

Mammillary Bodies

The mammillary bodies are two small ball-shaped structures that lie right next to one another just anterior to the midbrain; hence *Two peas in a pod that lie side by side*. They are part of the diencephalon, which comes from the Greek *Dia*, meaning between/across, and *enkephalos*, meaning brain. Thus, the diencephalon means *across-brain* or *between-brain*. The mammillary bodies are commonly affected by alcoholism, and subsequent thiamine deficiency can result in Wernicke's encephalopathy, which is characterized by ophthalmoparesis and ataxia, which can progress, albeit rarely, to the irreversible stage of Korsakov psychosis, which is characterized by anterograde amnesia and confabulation. Hence *With pints come lies and paralyzed eyes, And sideways you will lean.*

The mammillary bodies are part of the hypothalamus and consist of medial and lateral nuclei. The mammillary nuclei receive afferent information from the hippocampus via the postcommissural fornix. These nuclei also receive afferent information from the dorsal and ventral tegmental nuclei of Gudden in the midbrain via the mammillary peduncle. The efferent fibers from the mammillary nuclei are contained in fasciculus mammillaris princeps. This bundle splits into the mammillothalamic tract (destined for anterior thalamic nuclei) and the mammillotegmental tract of Gudden (back to the midbrain tegmentum).

The Papez circuit (hippocampus- > mammillary bodies- > anterior thalamus- > cingulate gyrus- > parahippocampal gyrus- > hippocampus) is considered to play an important role in memory formation. Hence, damage to mammillary bodies in thiamine deficiency is thought to be responsible for anterograde amnesia.

This is shown in Figure A.1.

3

• • • • •

Though some people call me funny,
I'm no bone, I'm here to say,
My roots are from two regions,
Without me your grip will fray.

Hint #1:

Think brachial plexus.

Hint #2:

Tingling/numb pinky.

Ulnar Nerve

The ulnar nerve is what is hit when people hit their elbow on an object, and it produces a tingling pain in their medial hand. Classically, this is called *hitting your funny bone*, but of course, it's *not a bone* at all.

The ulnar nerve is a peripheral nerve whose fibers travel through the brachial plexus (lower trunk and medial cord), originating from nerve roots C8–T1; hence *two regions* (*cervical* and *thoracic*). As the ulnar nerve descends through the arm, it does not give off any branches. It travels through the arcade of Struthers (formed by the medial head of triceps, internal brachial ligament, and deep fascia) and descends toward the elbow. It travels between the medial epicondyle of the humerus and the olecranon process of the ulna (retro-condylar groove). After this groove, the nerve enters the proximal forearm and travels under the humeral-ulnar aponeurosis (tendinous arch between the two heads of flexor carpi ulnaris), known as the cubital tunnel.

Within the forearm, the ulnar nerve gives off motor branches to the flexor capri ulnaris and the medial division of flexor digitorum profundus. Distally in the forearm, it gives off two sensory branches – the dorsal ulnar cutaneous and palmar ulnar cutaneous nerves. The dorsal ulnar cutaneous nerve supplies sensation to the dorsal medial hand and dorsal medial fourth and fifth fingers. The palmar ulnar cutaneous nerve supplies sensation to the palmar proximal medial hand (hypothenar region). The digital sensory branches that supply sensation to the palmar surface of medial fourth and fifth fingers arise distal to the wrist. Distal to the wrist, all the remaining motor branches arise. These provide motor function to most intrinsic hand muscles, including the dorsal and palmar interossei, adductor pollicis, ulnar lumbricals, and the deep head of the flexor pollicis brevis. Thus, if injured, your grip will be weak. Hence *your grip will fray.*

Ulnar neuropathy at the elbow can produce "ulnar claw hand" or "benediction posture" because of failure to extend the fourth and fifth digits (because of weakness of the lumbricals). It is associated with sensory loss in the distribution of dorsal, palmar, and digital cutaneous sensory branches. Ulnar neuropathy at or below the level of the wrist is associated with sensory loss only in the distribution of the digital sensory branches (palmar surface of medial fourth and fifth fingers).

4

• • • • •

From the middle to the muscles,
Through a fissure on my route,
Split five ways on my journey,
Without me you are down and out.

Oculomotor Nerve (CN III)

The oculomotor nerve, aka cranial nerve III, originates from its nucleus in the midbrain (*middle*), at the level of the superior colliculus, travels through the cavernous sinus, exiting the skull through the *superior orbital fissure*, and innervates the majority of the extraocular *muscles* (five in all: superior rectus, inferior rectus, medial rectus, inferior oblique, and levator palpebrae superioris). A complete lesion of the oculomotor nerve will produce weakness of eye elevation and adduction, thus leading to a resting position of depression and abduction. Hence *down and out*.

The oculomotor nuclear complex is in the midbrain at the level of superior colliculus. It is made up of multiple subnuclei, which give rise to fibers for corresponding muscles. Most of the fibers exit ipsilateral to the subnucleus and travel in the ipsilateral nerve, except for superior rectus. The fibers originating from the motor neurons in the superior rectus subnucleus cross within the nucleus and exit with the contralateral oculomotor nerve. The motor neurons for bilateral levator palpebrae muscles are in a single midline nucleus called the central caudal nucleus. After exiting the brainstem, the nerve is made up of two divisions – superior and inferior. The superior division supplies superior rectus and levator palpebrae superioris. The inferior division supplies the rest. It also carries the preganglionic parasympathetic fibers for the constrictor pupillae muscle.

An oculomotor nerve lesion will produce ipsilateral hypotropia, exotropia (down and out), ptosis, and mydriasis. A lesion of the oculomotor nucleus does not produce unilateral ptosis, unilateral mydriasis, or unilateral hypotropia.

5

• • • • •

A product of the "bulb" and called,
A "wanderer" on a quest,
Through a tunnel with two close friends,
To rest and digest.

Vagus Nerve (CN X)

The vagus nerve (CN X) originates in the medulla, which was archaically called the *bulb* (also where the commonly used term "bulbar" comes from), and also called the myelencephalon. The term *vagus* has Latin roots, meaning "wanderer/ing," and may have acquired its name due to its winding path through the body and subsequent innervation of many organs. It exits the skull through the jugular foramen (like a *tunnel*) with two other nerves – glossopharyngeal nerve (CN IX) and spinal accessary nerve (CN XI). Though the vagus nerve has many functions, one of its main functions is to provide parasympathetic innervation to the heart and gut, and thus help to *rest and digest.*

The preganglionic parasympathetic fibers for the heart, lungs, and gastrointestinal tract arise from the dorsal motor nucleus of vagus. The branchial motor fibers, which innervate pharyngeal and laryngeal muscles involved in swallowing and articulation, arise from the nucleus ambiguus. The viscerosensitive information from the heart, lungs, gastrointestinal tract, chemoreceptors, and baroreceptors is carried by the vagus nerve to the caudal part of nucleus tractus solitarius.

The main branch of the vagus nerve that controls the laryngeal muscles required for articulation is the recurrent laryngeal nerve that first descends into the thorax and then loops around the aortic arch to ascend toward the larynx. Voice hoarseness, dysarthria, and dysphagia are the main symptoms of vagus nerve injury.

Jugular foramen syndrome or Vernet syndrome is a constellation of symptoms, often produced by a skull-based tumor near the jugular foramen causing CN IX lesion (loss of taste sensation in posterior third of tongue), CN X lesion (hoarseness, swallowing difficulty, etc.), and CN XI lesion (weakness of ipsilateral shoulder elevation and head rotation toward the contralateral side).

6

• • • • •

Fed from all trunks and a favorite of drunks,
With roots from them all,
I split both deep and superficial,
Won't let your carpals fall.

Hint #1:

Brachioradialis is commonly called the "beer drinking muscle."

Hint #2:

Think brachial plexus.

13

Radial Nerve

The radial nerve contains *fibers from all three trunks* (superior, middle, and inferior), and *all five nerve roots* of the brachial plexus. It innervates brachioradialis muscle, which is known as the *beer-drinking muscle* as it functions to flex the arm at the elbow in a semi-pronated position, as when holding a beer.

Each trunk of the brachial plexus divides into an anterior and a posterior division. The posterior divisions of all trunks combine to form the posterior cord. Before the posterior cord continues as the radial nerve, it gives off axillary, thoracodorsal, and subscapular nerves as branches. In the arm, the radial nerve gives off the posterior cutaneous nerve of the arm, lower lateral cutaneous nerve of the arm, and posterior cutaneous nerve of the forearm. It also gives off branches to triceps before it enters the spiral groove of the humerus. In the distal arm, it gives off branches to brachioradialis and brachialis muscles. After crossing the elbow, the nerve splits into a *deep branch* and a *superficial sensory branch*. This sensory branch supplies sensation over the lateral part of the dorsal surface of the hand and dorsal surface of index, middle, and ring fingers.

The deep radial branch (aka posterior interosseous nerve) is a pure motor nerve, and it innervates all the extensor muscles of the wrist and the fingers. Hence, a lesion of either this branch or the proximal radial nerve will produce a wrist drop (*carpals fall*).

A lesion of the radial nerve at the level of spiral groove will result in wrist drop and weakness of the brachioradialis while sparing the triceps. It will also lead to sensory loss in the distribution of the superficial sensory branch. A lesion above the level of spiral groove (axilla) will also affect the triceps, causing sensory loss over the posterior forearm and arm. A lesion of the posterior interosseous nerve only produces wrist drop and spares both brachioradialis and triceps, causing no sensory loss.

Rarely, isolated superficial radial sensory neuropathy can develop secondary to compression of this nerve against the bone by tight bands, watches, or cuffs. This often presents as a painful syndrome affecting the dorsal hand known as cheiralgia paresthetica.

7

• • • • •

Twelve will move and tips will feel,
Through a tunnel near its end,
Painful in desk laborers,
When the carpals forward bend.

Hint #1:

Commonly entrapped.

Hint#2:

Phalen's test.

17

Median Nerve

Median nerve innervates *12 muscles and the tips of 3½ fingers.* On its way to the hand, it *passes through the carpal tunnel at the wrist.* It is *commonly injured in desk workers* due to compression, producing inflammation and resulting in damage found in carpal tunnel syndrome. One commonly utilized test for carpal tunnel syndrome and thus median nerve impingement is the reproduction of pain with Phalen's test, which is performed by *forward flexing the wrist* and pressing the dorsal aspects of the hands together.

The median nerve is formed by the combination of medial and lateral cords of the brachial plexus. The fibers from the medial cord are from C8–T1 roots and are mainly responsible for supplying the muscles. The fibers from the lateral cord are from C6–C7 roots and are mainly sensory fibers supplying the thumb, palmar, and dorsal aspects of index and middle fingers and the thenar eminence.

The anterior interosseous nerve (AIN) is the largest branch of the median nerve. It branches off the main trunk just distal to pronator teres. AIN syndrome is characterized by weakness of flexor pollicis longus, flexor digitorum profundus to digits 2 and 3, and pronator quadratus. Clinically, this is easily elicited by asking the patient to make an "OK" sign, which shows the inability to flex the distal thumb and index fingers.

Median neuropathy at the wrist (carpal tunnel syndrome), on the other hand, is mostly a sensory syndrome. Weakness of abductor pollicis brevis and opponens pollicis can be seen in advanced cases.

8

• • • • •

A child of the largest one,
That keeps your toes up high,
It hugs the neck then splits its trek,
Pinched when you are thigh on thigh.

Hint #1:

Some bones have anatomical "necks."

Hint #2:

Think "crossed legs."

Common Peroneal Nerve

The common peroneal (or fibular) nerve is a branch that *arises from the sciatic nerve* (the largest peripheral nerve in the body). The motor functions of the common peroneal nerve are *dorsiflexion at the ankle and toe extension* (via its deep fibular branch) and eversion at the ankle (via the superficial fibular branch). In its path, after branching laterally from the sciatic nerve, *it then hugs the neck of the fibula* before *splitting into the deep and superficial peroneal nerves*. Clinically, it is *commonly compressed when patients cross their legs,* thus applying a lateral pressure on the nerve as it passes over the fibular neck.

The common peroneal nerve arises primarily from L4-S1 nerve roots. The peroneal fibers and the tibial fibers of the sciatic nerve travel separately within the sciatic nerve. Within the common peroneal nerve, the deep peroneal fibers are more medial whereas the superficial peroneal fibers are more lateral.

Before the common peroneal nerve wraps around the neck of the fibula, it gives off a sensory branch called the lateral cutaneous nerve of the knee, which supplies sensation to the lateral aspect of the knee.

The deep peroneal nerve innervates the muscles involved in dorsiflexion of the ankle and the toes – tibialis anterior, extensor digitorum longus, extensor digitorum brevis, and extensor hallucis longus. The superficial peroneal nerve innervates ankle evertors — peroneus longus and brevis, and it also provides sensation to a majority of the lateral calf.

A common peroneal neuropathy at the fibular neck presents with foot drop and steppage gait with sensory loss in the dorsum of foot and lateral calf. A common peroneal neuropathy above the fibular neck within the sciatic nerve, in addition, will also have sensory loss on the lateral knee surface. On electrodiagnostic testing, changes are seen in the short head of biceps femoris as well because this muscle is supplied by the peroneal fibers of the sciatic nerve.

Sciatic neuropathy can be easily distinguished from common peroneal neuropathy by examining muscles supplied by the tibial branch of the sciatic nerve – plantarflexors and ankle invertors.

A major differential for common peroneal neuropathy is an L5 radiculopathy. These can be differentiated clinically by testing foot inversion, which is a function of muscles innervated by the tibial nerve. In addition, the weakness of toe dorsiflexion (extensor hallucis longus) is out of proportion to the weakness of ankle dorsiflexion (tibialis anterior) in L5 radiculopathy as opposed to common peroneal neuropathy.

9

• • • • •

The largest and the longest,
With compressive pear-shaped friend,
And rear-thigh pain that radiates,
Thanks to me your knees will bend.

Hint #1:

Think radiating low back pain.

Hint #2:

Piriformis is Latin for "pear-shaped."

Sciatic Nerve

The sciatic nerve is the *largest and longest nerve in the body*. After being formed by nerve roots L4–S3, it travels infero-laterally through the gluteal area, *passing under the piriformis muscle* (Latin for "pear-shaped"), which is often a site of *compression*. Pain often occurs in the proximal thigh with radiation into the leg posteriorly. There is usually no back pain. "Sciatica" is referred to as a *clinical syndrome of back pain radiating into the thigh and leg*. This is a radiculopathy (not sciatic neuropathy) caused by the impingement of one of the lumbo-sacral nerve roots. In that sense, the term "sciatica" is a misnomer.

The sciatic nerve is made up of tibial and peroneal fibers. Above the level of the fibular neck, the sciatic nerve innervates medial and lateral hamstrings and the lateral division of adductor magnus. The short head of the biceps femoris is the only "peroneal" innervated muscle. The remaining muscles are innervated by the tibial fibers. The primary motor functions of the sciatic nerve include *knee flexion*, which is the function of the hamstrings.

Sciatic neuropathy can clinically mimic a lower lumbosacral plexopathy or L4–5/S1 radiculopathy. These can be differentiated on exam and electrodiagnostic testing by examining the muscles supplied by the superior gluteal nerve, which has the same root value. The two important muscles supplied by the superior gluteal nerve are gluteus medius and tensor fasciae latae (TFL). If these muscles are weak and/or have evidence of denervation on electro-diagnostic testing, then the lesion is likely to be proximal to the sciatic nerve (either lower lumbosacral plexus or a radiculopathy).

10

● ● ● ● ●

Sharp turn and through a tunnel,
With close friends both red and blue,
Though I start above the fossa,
Push down your toes is what I do.

Tibial Nerve

The tibial nerve is a branch of the sciatic nerve. It *branches off superior to the popliteal fossa,* and descends toward the leg anterior to reach the calf muscles. It makes a *sharp turn after passing the ankle, traveling through the tarsal tunnel* along with the *posterior tibial artery and vein (red and blue).* Primary motor function of the tibial nerve includes plantar flexion, *toe flexion,* and inversion at the ankle.

Proximally, in the thigh, all the hamstrings (except for the short head of biceps femoris) are innervated by the tibial compartment of the sciatic nerve.

Foot inversion is an important clinical test to differentiate common fibular neuropathy from a more proximal lesion (e.g., sciatic neuropathy or L5 radiculopathy). This is because foot inversion is controlled by the tibial nerve with the same root value of L5 as that of foot dorsiflexion and eversion, which are controlled by the common fibular nerve.

The tarsal tunnel is formed by the flexor retinaculum and the medial malleolus. As the nerve passes under it with the tibial artery and vein, it gives off the medial calcaneal sensory nerve and then medial and lateral plantar nerves. The most common complaint in tarsal tunnel syndrome is pain around the malleolus.

11

• • • • •

Though there are only two of twelve,
Outside the lizard brain reposed,
Continues through the crib of ethmos,
It is the only one exposed.

Olfactory Nerve (CN I)

Cranial nerve one, the olfactory nerve, is *one of the only two nerves (CN I and II) that do not have a nucleus in the brainstem*, also known as *lizard brain*. Near its endpoint, fibers from the olfactory nerve travel through the *cribriform plate, which is a region of ethmoid bone*. It is unique because it is the only nerve that is *exposed to the outside environment*.

After synapsing in the olfactory bulb, the information is further conducted via the olfactory tract, which runs in the olfactory sulcus (aka medial orbital sulcus) in the base of the frontal lobe. The olfactory tract is made up of axons of mitral and tufted cells. These axons travel directly to the olfactory cortex without a thalamic relay. Some of these axons send collaterals to the neurons in the anterior olfactory nucleus, which in turn sends axons to both ipsilateral and contralateral olfactory bulbs via the anterior commissure.

The primary olfactory cortex is located in the piriform cortex and the periamygdaloid cortex around the amygdala. Some fibers of the olfactory cortex do indeed synapse with the corticomedial nucleus of the amygdala.

The basal part of the frontal lobe, where the olfactory tract travels, is quite susceptible to injury sustained during head trauma because of its proximity to the bony skull base. This makes the olfactory tracts more susceptible to injury as well, hence making hyposmia or anosmia a common complaint in patients with moderate to severe traumatic brain injury (TBI).

In cases of refractory post-traumatic epilepsy with poor or no electroencephalogram (EEG) correlate, anosmia can be a hint that the suspected epileptogenic zone is localized within the base of the frontal lobe rather than the temporal lobe.

12

• • • • •

From the dark side of the moon,
Through a canal and draped in mother,
Special sense through intersection,
Lights off to one opens the other.

Optic Nerve (CN II)

The optic nerve (formed by a collection of axons of retinal ganglion cells) originates from the retina, which is in the *back of the eye* (dark side of the moon), traveling posteriorly *through the optic canal*, where it is *covered completely by dura mater (dura mater is Latin for "tough mother")*. It functions to *transmit sight* (special sense) from the eye to the brain, *crossing paths at the optic chiasm*, allowing fibers from each optic nerve to reach both sides of the brain. It also plays an important role in the *pupillary light reflex*, during which both pupils will constrict or dilate in response to a light or dark environment, respectively.

The optic nerve exits the orbit via optic canal located at the orbital apex next to the superior orbital fissure. Within the optic nerve, the axons are arranged in a retinotopic manner – nasal fibers (temporal visual field) and temporal fibers (nasal visual field). Similarly, the inferiorly located fibers correspond to the superior half of the visual field and superiorly located fibers correspond to the inferior half of the visual field. At the level of the chiasm, only the nasal fibers decussate, whereas the temporal fibers continue in the ipsilateral tract.

A lesion of the optic nerve aka optic neuropathy (prechiasmal) will lead to visual disturbances in the ipsilateral eye only. A lesion at the level of the chiasm produces bitemporal hemianopia (because it affects the nasal retinal fibers crossing in the middle of the chiasm). A lesion of the optic tract (retrochiasmal) leads to contralateral homonymous hemianopia (because it affects temporal fibers from the ipsilateral eye and nasal fibers from the contralateral eye). Rarely, a tumor at the junction of the optic nerve and the chiasm can produce ipsilateral vision loss by the compression of the optic nerve and contralateral superior temporal quadrantanopia, known as junctional scotoma. Even more rarely, the tumor can only compress the nasal fibers of the ipsilateral optic nerve, thus producing an ipsilateral temporal scotoma known as junction scotoma of Traquir.

13

• • • • •

The largest of the twelve siblings,
Split three ways from central hub,
Of my many unique purposes,
A lover's touch and crushing grub.

Hint #1:

Has a very large nucleus, aka, central hub.

Hint #2:

Largest, but not the longest.

Trigeminal Nerve (CN V)

The trigeminal nerve is the *largest of the twelve cranial nerves*. After traveling anterior from its origin in the pons to the trigeminal ganglion (*central hub*), it splits into *three main branches*: ophthalmic (V1), maxillary (V2), and mandibular (V3), which provide sensation to the face (*where you would feel a lover's touch*). The nerve also provides motor innervation to the muscles of mastication, thus *helping with chewing (crushing grub)*.

The trigeminal nerve has three nuclei – chief sensory nucleus, mesencephalic nucleus, and spinal trigeminal nucleus. The chief sensory nucleus receives afferent information from all branches for fine touch and pressure. The mesencephalic nucleus primarily receives proprioceptive input. Pain and temperature go to the spinal trigeminal nucleus.

The V1 division enters via the superior orbital fissure along with CN III, IV, and VI. These nerves travel in the cavernous sinus along with the V2 division, which enters via the foramen rotundum. After the cavernous sinus, these are joined by the V3 division, which enters via foramen ovale. All the nerves join to form the gasserian ganglion or trigeminal ganglion in Meckel's cave. After synapsing with second-order neurons in the chief and the spinal nuclei, the axons cross the midline and ascend as trigeminal lemniscus and trigeminothalamic tract, respectively.

The mesencephalic trigeminal nucleus, which receives proprioceptive input from muscles of mastication contains the primary neurons within the nucleus rather than in a ganglion outside the central nervous system. This nucleus mediates the jaw jerk.

A lesion of the trigeminal nerve either within the pons or the pre-pontine cistern will lead to loss of sensation in all three distributions (V1, V2, and V3). A lesion of the trigeminal nerve within the cavernous sinus will only lead to sensory loss in V1 and V2 distributions (V3 splits away from the main trunk before the cavernous sinus and exits the skull base via foramen ovale). A lesion of the trigeminal nerve in the superior orbital fissure (part of orbital apex) will only produce sensory loss in V1 distribution (V2 splits away after exiting from cavernous sinus via foramen rotundum).

Numb chin syndrome (aka mental neuropathy) is an important clinical sign characterized by isolated numbness in the chin along the distribution of the mental nerve (a terminal branch of V3 division). In addition to local structural causes, this condition is often related to systemic malignancies. Whole body PET/CT should be ordered in these patients.

14

• • • • •

Exiting beneath the bridge,
Through cave after Gruber it passes,
Though it serves only one purpose,
Lesion it, wear prismed glasses.

Hint #1:

Latin word for "bridge" is *pontem/pons*.

Hint #2:

Gruber's ligament (aka petroclinoid or petrosphenoidal ligament).

Abducens Nerve (CN VI)

The abducens nerve originates in the pons (*pons* means "bridge" in Latin). The abducens nucleus is located in the caudal dorsomedial pons. The fibers travel ventrally through the corticospinal tract and exit at the ponto-medullary junction. The nerve then travels in the pre-pontine cistern and ascends toward the petroclival confluence. This venous confluence is separated into a superior and inferior compartment by the Gruber's ligament (also known as petroclinoid ligament). The abducens nerve travels in the inferior compartment where it is tethered to the *Gruber's ligament*, known as Dorello's canal. Afterward, the nerve enters the *cavernous sinus*. In the cavernous sinus, the nerve leaves its dural sleeve and runs very close to the internal carotid artery. The nerve exits the cavernous sinus with CN III, IV, and V1 via the superior orbital fissure to enter the orbit. In the orbit, it runs next to the superior ophthalmic vein and *only innervates one muscle*, the lateral rectus.

Weakness of the lateral rectus produces abduction paresis of the affected eye. One treatment for the subsequent horizontal diplopia is *wearing glasses with prism-corrected lenses*.

A lesion of the abducens nucleus itself in the pons produces complete horizontal gaze palsy toward the ipsilateral side. But a lesion of the abducens nerve only leads to abduction paresis of the ipsilateral eye.

A lesion of the abducens nucleus along with medial longitudinal fasciculus (MLF) leads to the so-called one and a half syndrome. This is characterized by complete horizontal conjugate gaze palsy toward the ipsilateral side and adduction paresis of the ipsilateral eye on attempted contralateral conjugate gaze (aka internuclear ophthalmoplegia).

15

• • • • •

With five branches, many tributes,
Two of which go toward the ear,
Through temporal bone it travels,
Express, and taste, and shed a tear.

Hint #1:

A tribute of this nerve, protects the inner ear from LOUD NOISES!!!

Hint #2:

Also involved in salivation.

Facial Nerve (CN VII)

The facial nerve is a mixed nerve with motor, visceral sensory, somatic sensory, and parasympathetic functions. The motor facial nucleus is located in caudal dorsal pons at the level of abducens nucleus. As the fibers emerge, they wrap around the abducens nucleus, forming the so-called facial colliculus. As the nerve exits the brainstem, it is accompanied by a smaller nerve called nervus intermedius. The main nerve carries the motor fibers from the facial nucleus. The nervus intermedius carries the sensory fibers and the preganglionic parasympathetic fibers.

The parasympathetic fibers arise from the superior salivatory nucleus. The somatic sensory fibers end up in the spinal trigeminal nucleus, whereas the visceral sensory fibers end up in the rostral nucleus tractus solitarius.

The facial nerve and the nervus intermedius enter the internal auditory meatus and travel with the vestibulocochlear nerve. At the genu, the nerve takes a posterior turn. The geniculate ganglion in the genu contains the primary sensory neurons. Some of the parasympathetic fibers split here from the main nerve and travel as greater petrosal nerve in the vidian canal before reaching the sphenopalatine ganglion. Postganglionic fibers from this ganglion travel with the maxillary nerve to reach the lacrimal glands. The rest of the nerve travels in the facial canal. Chorda tympani arises from this trunk and carries parasympathetic fibers for submandibular ganglion and taste fibers to the anterior two-thirds of the tongue.

The rest of the nerve exits the skull through stylomastoid foramen and splits into *five main branches* (frontal/temporal, zygomatic, buccal, marginal mandibular, and cervical). Major functions of the facial nerve include *tear production via the lacrimal glands*, *nerve to stapedius muscle (loss of function leading to hyperacusis)*, *taste to anterior two-thirds of tongue* and *salivation via the submandibular gland,* and innervating the muscles of *facial expression.*

A peripheral lesion of the facial nerve proximal to the geniculate ganglion will lead to deficits in all the functions listed above. A lesion distal to the geniculate ganglion but proximal to the origin of nerve to stapedius will spare the lacrimation. A lesion distal to the origin of the nerve to stapedius and proximal to the origin of chorda tympani will not present with hyperacusis. And finally, a lesion distal to the origin of chorda tympani will only present with facial weakness.

16

· · · · ·

From the bridge as one, from two,
Split one to snail and one to shell,
To not only remain upright,
But appreciate the dinner bell.

Hint #1:

Think about the general structure and shape of a snail and snail shell.

Hint #2:

Possesses two distinct components.

Vestibulocochlear Nerve (CN VIII)

The vestibulocochlear nerve exits the pons as one nerve; however, the vestibular components come from nuclei in the pons and medulla while the cochlear components come from nuclei in the inferior cerebellar peduncle, thus the nerve enters the pons (derived from the Latin word for "bridge") as two components but exits the pons as one combined nerve with both vestibular and cochlear components. After merging and traveling through the internal acoustic meatus, the two components split on exit, with the vestibular component/vestibular nerve heading to the otolith organs and semicircular canals (*that look like a snail*) and the cochlear components/cochlear nerve going to the cochlea (*shell*). The primary functions of these nerves include regulating balance (vestibular nerve) and hearing (cochlear nerve).

Neurological causes of unilateral hearing loss include disorders of the cochlea itself, the cochlear nerve, or the cochlear nuclei. Other lesions beyond the level of cochlear nuclei do not produce unilateral hearing loss as the auditory pathway has bilateral projections at every level after cochlear nuclei.

17

• • • • •

Coming from an outside street,
A short distance I will go,
Fed by, and to feed, only three,
Your ticket to the gun show.

Hint #1:

"outside"... think, lateral.

Hint #2:

When kids show off their muscles.

Musculocutaneous Nerve

The musculocutaneous nerve is a terminal branch of the *lateral cord of the brachial plexus.* Uniquely, it *travels a relatively short distance compared to other branches* of the brachial plexus. It includes fibers from only *three nerve roots* (C5, C6, and C7) and innervates *only three muscles* (biceps, brachialis, and coracobrachialis). The major function of the musculocutaneous nerve is to flex the elbow, *thus flexing the biceps muscle* (showing others your "guns").

After the elbow, the nerve ends as a pure sensory nerve called the lateral antebrachial cutaneous sensory nerve, which supplies sensation to the lateral half of the forearm.

Isolated musculocutaneous neuropathy is rare. It has been reported to occur in people who frequently carry items on their shoulder with the arm curled around the object aka "carpet carrier's palsy." In most cases, the entrapment of the nerve occurs distally between the biceps tendon and the brachialis muscle. This leads to an isolated sensory loss over the lateral forearm.

18

· · · · ·

Helpful when things below get tachy,
Massage and I'll decrease the rate,
Though I'm often perceived bitterly,
I also make you salivate.

Hint #1:

Think parasympathetics.

Hint #2:

I *sense* a gag coming.

Glossopharyngeal Nerve (CN IX)

Glossopharyngeal nerve is a mixed cranial nerve that contains para-sympathetic fibers, which innervate the carotid sinus/body, which *if massaged will decrease the heart rate*. Additionally, it provides special sense (*taste) to the posterior third of the tongue (bitter area)* as well as innervating the parotid glands, which *helps with salivation*. It also functions as the afferent/sensory arm of the gag reflex.

The preganglionic parasympathetic fibers arise in the inferior salivatory nucleus. These fibers end up in the otic ganglion via the lesser petrosal nerve. Postganglionic fibers end up in the parotid gland.

The visceral sensory fibers (taste, chemoreceptor, and baroreceptor) end up in the nucleus tractus solitarius (rostral and caudal, respectively).

The nerve only supplies one muscle called the stylopharyngeus, and the motor neurons are located in the nucleus ambiguus. The nerve exits the skull via the jugular foramen.

Glossopharyngeal neuralgia (GN) is a rare craniofacial pain syndrome characterized by recurrent lancinating attacks of pain in the posterior part of the tongue, pharynx, and/or in the ear. The pain can radiate to the submandibular region as well. It is often precipitated by coughing, yawning, swallowing, or talking.

Rarely, patients with GN can present with bradycardia, hypotension, and transient asystole leading to recurrent episodes of cardiac syncope. The mechanism is likely that the visceral afferent impulses conducted by the glossopharyngeal nerve reaching the nucleus tractus solitarius could reach the dorsal motor nucleus of vagus nerve via collaterals, thus causing the bradycardia. The other possible mechanism is that the Hering's nerve (aka carotid sinus nerve), which is a branch of the glossopharyngeal nerve and terminates in the dorsal motor nucleus of the vagus nerve, provides a path for sensory afferents to activate the parasympathetic fibers destined for the heart.

19

• • • • •

I first go up, and then come down,
Through two foramen, one with heft,
Though I go to only two places,
Use your right and you'll look left.

Hint #1:

Innervates two muscles ("to only two places").

Hint #2:

Bracelets, necklaces, cufflinks, earrings are types of . . . .

Spinal Accessory Nerve (CN XI)

The spinal accessory nerve is a unique cranial nerve that originates in the spinal cord and *ascends up through the foramen magnum (the one with heft) then descends back through the jugular foramen.* It only *innervates two muscles* (sternocleidomastoid [SCM] and trapezius).

Because of the origin and insertion of the sternocleidomastoid muscles, *contracting one side will cause head rotation to the opposite side.* Thus, if right SCM is activated, the head will turn left. The trapezius elevates the shoulder.

With hemispheric lesions involving the motor cortex like strokes, the patients develop dissociated weakness of the ipsilateral SCM and contralateral trapezius muscles. Brainstem lesions below the ponto-mesencephalic junction and above the cervico-medullary junction present with contralateral weakness of trapezius and SCM instead.

20

. . . . .

Exiting from bottom third,
Through a hole that shares my name,
If use your tongue to catch snowflakes,
Right lesions mean rightward you'll aim.

Hint #1:

Can narrow it down by first naming the cranial foramen.

Hint #2:

Google image search "catching snowflakes."

Hypoglossal Nerve (CN XII)

The hypoglossal nerve is a pure motor nerve with its nucleus located in the medial caudal medulla. The nerve *exits from the medullary portion of the brainstem (bottom third)* and travels *through the hypoglossal canal (a hole that shares my name).* It is commonly tested on neurological exam by asking patients to stick out their tongue and looking for deviation. Thus, if you are trying to catch a falling snowflake on your tongue, a *lesion of CN XII will cause ipsilateral tongue deviation.*

Acute ischemic stroke involving the motor cortex can sometimes present with isolated tongue weakness (aka "cortical tongue") because of large motor representation of the tongue. In this case, the tongue will deviate away from the side of the lesion. On the other hand, tongue weakness caused by medial medullary infarcts (Dejerine syndrome) affecting the hypoglossal nucleus will cause the tongue to deviate toward the side of the lesion.

Collet–Sicard syndrome is a combination of CN IX, X, XI, and XII palsies, caused by lesions of the skull base involving both the jugular foramen and the hypoglossal canal.

An unusual combination of CN VI and CN XII palsies known as Godtfredsen syndrome is typically caused by a clival chordoma.

21

· · · · ·

Surnamed in Egypt, with walls tough to decrypt,
I switch sides, then head toward ground,
With lesion comes clonia, and matched hypertonia,
And time often spent wheelchair-bound.

Hint #1:

Think of the stereotypic symbol/shape of Egypt.

Hint #2:

You've got this. You're on the right tract.

Corticospinal Tract

Admittedly, this is probably the toughest riddle to get. First you must know that the corticospinal tract is *also called the pyramidal tract*. Thus, it would have received its surname in Egypt. This is also hinted in the reference to something with walls that are tough to decrypt (*hieroglyphics) being a key feature of pyramids*. The pyramidal tract/corticospinal tract is a key descending pathway that *decussates (switches sides) at the level of caudal medulla and descends toward the spinal cord* (heads toward ground). Lesions affecting the corticospinal tract will *produce* clonus and spastic paralysis (*clonia and hypertonia*). Many of those with said lesions will sadly remain *wheelchair-bound because of their injury*.

Most corticospinal fibers originate in the primary motor cortex (Brodmann areas 4 and 6) in the precentral gyrus and decussate at the caudal medullary level and descend in the ventrolateral funiculus of the spinal cord as the lateral corticospinal tract. A small percentage of fibers do not decussate and descend ipsilaterally as the anterior corticospinal tract.

At the origin, the leg fibers are medial since the leg is represented medially in the motor homunculus. As the corticospinal fibers descend through the corona radiata toward the posterior limb of the internal capsule, the fibers rotate in a way that the leg fibers become more lateral and posterior in the internal capsule and the arm fibers become more medial and anterior. This somatotopy of fibers is maintained in the cerebral peduncle and then in the spinal cord as well.

22

• • • • •

First I will cross, then I'll ascend,
A laterally placed band,
Having three within my chain,
I'm how you know you burnt your hand.

Hint #1:

Ouch!! That's hot!!

Hint #2:

Flexor withdrawal reflex primary afferents.

Spinothalamic Tract

The spinothalamic tract is an ascending pathway that *carries pain and temperature sensation*, that is, how you would know you burnt your hand. The first-order sensory neuron located in the dorsal root ganglion receives sensory information mainly via Aα and C fibers. The axons from the first-order neuron enter the spinal cord via the dorsal root. As opposed to the dorsal column pathway, these axons synapse immediately in laminae I and V of the dorsal root gray matter (substantia gelationsa). The axons from the second-order neurons then *immediately cross to the contralateral side of the spinal cord* through the central gray matter and *ascend via a lateral tract in the spinal cord* to synapse in the contralateral thalamus. Some of the axons first ascend or descend via Lissauer's tract before crossing to the contralateral side.

The third-order neurons, located in the ventral posterolateral nucleus (VPL) of the contralateral thalamus in turn project to the cortex.

The primary cortical target of the thalamic neurons relaying pain and temperature sensation is posterior insula. Some of the information is also relayed to the anterior cingulate gyrus.

A hemisection of the spinal cord (Brown–Sequard syndrome) leads to the ipsilateral loss of fine touch, proprioception, and vibration sensations and a contralateral loss of pain and temperature sensations.

23

• • • • •

First I ascend, and with sensation,
Two stops, pre-cross, both "wedge" and "slim,"
I continue up the other side,
Feel vibrating/positioned limb.

Hint #1:

Latin word for "wedge" is *cuneatus* and Latin word for "slim" is *gracilis*.

Hint #2:

A three-step process.

Dorsal Columns

The dorsal column medial lemniscus pathway is responsible for *transmission of vibration and position sense*. As one of the *key ascending sensory pathways*, it brings sensory information up the spinal cord from the periphery.

The first-order sensory neuron is located outside the spinal cord in the dorsal root ganglion. The axons carrying fine touch, vibration, and proprioception are mostly of large diameter and are made up of myelinated Aα and Aβ fibers. As the axons enter the spinal cord through the dorsal root, they enter the dorsal or posterior funiculus of the spinal cord and ascend without synapsing, as *fasciculus gracilis* (*gracile* means thin in Latin) and *fasciculus cuneatus* (*cuneate* means wedge in Latin). The *fasciculus gracilis* is more medial and represents leg fibers. The *fasciculus cuneatus* is more lateral and represents upper trunk and arm fibers.

These axons synapse with second-order neurons in the nucleus gracilis and nucleus cuneatus, respectively, at the level of caudal medulla. After *synapsing, fibers then immediately decussate* as internal arcuate fibers and then *ascend in the contralateral medial lemniscus* to the contralateral thalamus. In the thalamus, the fibers synapse with the third-order neurons in the VPL nucleus. These neurons, in turn, project to the primary somatosensory cortex in the postcentral gyrus.

24

• • • • •

The first two of four cavum,
And also the most grand,
I carry *liquor cotunnii*,
Agaped, C-shaped, and if dammed, expand.

Hint #1:

Liquor cotunnii is a non-blood bodily fluid.

Hint #2:

Come as a pair.

Lateral Ventricles

The ventricular system represents a network of cavities within the brain. The ventricles contain cerebrospinal fluid, which was previously named *liquor cotunnii,* after the famous eighteenth-century Italian physician and anatomist Domenico Cotugno. Cerebrospinal fluid is produced by the choroid plexus.

The lateral ventricles are the largest two of the four ventricles that cerebrospinal fluid flows through. These are C-shaped on sagittal view, and expand in the case of hydrocephalus.

The lateral ventricles have three horns – frontal, temporal, and occipital. The frontal horn is anterior to the foramen of Monro (which connects the lateral ventricle to the third ventricle). Posterior to the frontal horn is the body and then the atrium, which is in turn connected to the occipital horn and the temporal horn.

Both ventricles (including their horns) typically appear as narrow slits on CT or MRI. In patients with impending hydrocephalus, one of the earliest signs is the dilation of the temporal horns of the lateral ventricles.

This is shown in Figures A.1 and A.2.

25

• • • • •

I am one of the four great lakes,
The first of two that sit midline,
Though we are all four fed from plexus,
I sit afront the gland of pine.

Hint #1:

Stay away from the brachial plexus.

Hint #2:

Third time's the charm.

Third Ventricle

The third ventricle is one of the four cerebrospinal fluid-filled ventricles and is the first midline ventricle into which cerebrospinal fluid flows from the lateral ventricles, and from which cerebrospinal fluid flows to the fourth ventricle. Though all four ventricles have a choroid plexus that help create cerebrospinal fluid, the third ventricle is the only one whose posterior wall is comprised of the pineal gland (named for its resemblance to the pine cone). The lateral ventricles drain into the third ventricle via the foramen of Monro. The third ventricle, in turn, drains via the cerebral aqueduct of Sylvius into the fourth ventricle. The thalamus and hypothalamus form the walls of the third ventricle.

A commonly used surgical procedure to treat children with obstructive hydrocephalus caused by conditions such as aqueductal stenosis is endoscopic third ventriculostomy (ETV). This technique involves introducing an endoscope via a burr hole from the vertex through the foramen of Monro into the third ventricle. Both the foramen of Monro and the third ventricle have to be of sufficient width to allow the endoscope to pass through them (>7 mm). A blunt fenestration of the floor of the third ventricle is made to provide an alternative passage for the cerebrospinal fluid (CSF) to drain.

This is shown in Figures A.1 and A.2.

26

• • • • •

A canal pos'd just prior to four,
And compared to it, I'm narrow more,
From north to south, within this nexus,
I am without productive plexus.

Hint #1:

Think marvel of Roman engineering (inventor: Appius Claudius 312 BC).

Hint #2:

I have four humps on my posterior wall.

Cerebral Aqueduct

The cerebral aqueduct is a canal that bridges the cerebrospinal fluid flow between the third and fourth ventricles of the brain. It is very narrow compared to these adjacent ventricles. Additionally, it is the only large major structure in the cerebrospinal fluid flow pathway that does not have choroid plexus. It is positioned just anterior to the quadrigeminal plate of the midbrain (made up of superior and inferior colliculi).

Congenital aqueductal stenosis is a common cause of obstructive hydrocephalus in children. Pineal gland tumors such as germinoma often present with compression of the aqueduct, thus leading to obstructive hydrocephalus. These patients have characteristic signs and symptoms, including failure to elevate their eyes (supranuclear upgaze paralysis), retraction of both eyelids (Collier's sign), convergence retraction nystagmus, and light near dissociation (failure of light reflex but preserved near response). This constellation of signs and symptoms is called Parinaud syndrome or dorsal midbrain syndrome.

This is shown in Figure A.1.

27

• • • • •

Though I'm the last of the four lochs,
With apertures both mid and lat,
If compressive force, my flow blocks,
You'll headache and outsize your hat.

Hint #1:

Put your hand in the air, like you just do not Chiari.

Hint #2:

Many call it a "diamond-shaped" cavity.

Fourth Ventricle

The fourth ventricle is the *last of the four ventricles* (which are like cerebrospinal lakes, *lochs*) through which the cerebrospinal fluid flows. It contains the outflow tracts/apertures foramen of Magendie (aka *medial aperture*) and foramen of Luschka (aka *lateral aperture*). It can be the focus of compression from tumors, or abscesses, or Chiari malformation, which *causes obstruction of cerebrospinal fluid flow*, producing hydrocephalus, which in turn commonly causes *headache and increased cranial size in kids.*

An important structure in the floor of the fourth ventricle is called the area postrema. These are paired vascular structures in the medulla, adjacent to the nucleus of tractus solitarius and facing the foramen of Magendie. The area postrema is one of the many circumventricular organs (which lack blood–brain barriers) and can detect chemical messengers in the blood. It is the chemoreceptor trigger zone and plays an important role in vomiting.

A clinical condition known as area postrema syndrome consists of intractable nausea, vomiting, and/or hiccups. This is often seen in patients with neuromyelitis optica spectrum disorder (NMOSD) characterized by aquaporin 4 (AQP4) antibodies.

This is shown in Figure A.1.

28

• • • • •

Covered over by a tent,
Sometimes obstructive with decent,
Balance and posture when you train,
I'm the Latin little brain.

Hint #1:

Not the word for "pre-Civil War" but close.

Hint #2:

10% of the volume, >70% of the total neurons.

Cerebellum

The cerebellum is covered by a dural reflection called the *tentorium cerebelli,* aka the *tent.* If the cerebellum descends into the foramen magnum, it *can cause obstructive hydrocephalus* by obstructing the flow of the cerebrospinal fluid. It functions to *help with balance and posture* and gets its name from Latin, meaning *little brain.* Though the cerebellum takes up only approximately 10% of the brain's volume, it contains over 70% of its total neurons.

The cerebellum consists of two hemispheres, a midline vermis and a small flocculonodular lobe. The vermis mostly works to control axial muscles, maintaining posture, and the hemispheres are associated with appendicular muscle coordination. The flocculonodular lobe is mainly involved in controlling eye movements.

The cerebellum is connected to the brainstem via superior, middle, and inferior cerebellar peduncles. The superior cerebellar peduncle (midbrain) is the output pathway of the cerebellum. Middle (pons) and inferior (medulla) cerebellar peduncles are input pathways carrying information via mossy and climbing fibers, respectively. These input fibers synapse on the Purkinje cells in the cerebellar cortex, which in turn project onto the deep cerebellar nuclei (dentate, emboliform, globose, and fastigial). The deep cerebellar nuclei then project out via the superior cerebellar peduncle.

Vermian atrophy, as seen in alcoholics, presents with truncal ataxia and falls. Hemispheric lesions, such as strokes, present with ipsilateral appendicular ataxia, which can be detected by dysmetria on finger to nose and heel to shin testing. Flocculonodular lobe lesions present with saccadic dysmetria and spontaneous nystagmus. The most typical example is the multidirectional gaze-evoked nystagmus with rebound phenomenon.

This is shown in Figure A.1.

29

• • • • •

Planning, movement, late evolving,
Language production, problem solving,
A scented bulb I sit atop,
Thanks to me, know when to stop.

Hint #1:

Release signs.

Hint #2:

Largest of its kind.

Frontal Lobe

The frontal lobe contains the prefrontal cortex, which *aids with planning and problem solving*, and *matures/evolves late in development*. It also contains *Broca's area*, which is a language area and is predominantly responsible for speech production. It is positioned just *above the olfactory bulb*, which perceives scent. Interestingly it also functions to *help with impulse control*, thus *knowing when to stop.*

The frontal lobe consists of a mesial surface (mesial frontal lobe and cingulate gyrus), a dorsolateral convexity (primary motor cortex, lateral premotor cortex, and dorsolateral prefrontal cortex), and a basal surface (orbitofrontal region). The frontal lobe is separated from the parietal lobe by the central sulcus. The sylvian fissure separates it from the temporal lobe. Most of the mesial and superior surface of the frontal lobe is nourished by the anterior cerebral artery, whereas most of the lateral convexity and basal region is supplied by the middle cerebral artery.

The most posterior part of the frontal lobe, which is immediately in front of the central sulcus, known as the precentral gyrus, contains Brodmann areas 4 and 6, which represent the primary motor cortex (M1). This area contains a representation of the body known as the Penfield homunculus. The leg and foot are represented on the mesial surface, whereas the face and hand are represented on the lateral surface.

All the frontal lobe regions perform different functions ranging from planning and execution of movements to executive functions and impulse control. Lesions of the orbitofrontal region are typically associated with impulse-control disorders and disinhibited behavior. A well-known historical example is the case of Phineas Gage, who had a dramatic personality change after an accident where a tamping iron rod went through his lower jaw into the skull, through his left frontal lobe, while he was working on the Rutland and Burlington Railroad in Vermont.

30

• • • • •

Make sense of words and help with math,
The end point of two upward paths,
Process vibration, touch, position,
Lesion right, full left omission.

Hint #1:

Do not neglect the right answer.

Hint #2:

Middle cerebral artery (MCA) territory.

Parietal Lobe

The parietal lobe is located posterior to the frontal lobe, anterior to the occipital lobe, and superior to the temporal lobe. The central sulcus separates the parietal lobe from the frontal lobe. Inferiorly, the sylvian fissure separates the parietal lobe from the temporal lobe anteriorly. Posteriorly, the distinction between parietal lobe and temporal lobe is less clear. The parieto-occipital sulcus separates the parietal lobe from the occipital lobe. The mesial part of the parietal lobe is called the precuneus. The lateral surface is split into the superior parietal lobule and the inferior parietal lobule by the intraparietal sulcus.

The postcentral gyrus (between the central sulcus and the postcentral sulcus) contains Brodmann areas 3, 1, and 2. These areas constitute the primary somatosensory cortex (S1). The superior parietal lobule contains Brodmann areas 5 and 7, whereas the inferior parietal lobule contains Brodmann areas 40 and 39. Area 40 is the supramarginal gyrus and area 39 is the angular gyrus. In the dominant hemisphere, supramarginal and angular gyri are part of the eloquent posterior language cortex (Wernicke's area).

An intriguing clinical syndrome called Gerstmann syndrome is caused by a lesion of the dominant angular gyrus. The typical findings of the Gerstmann syndrome are agraphia (inability to write), acalculia (inability to perform calculations), finger agnosia (inability to discriminate between the fingers), and right–left disorientation. This syndrome is a good example of the complex functions performed by the parietal lobe.

Lesions of the nondominant parietal lobe usually produce hemi-spatial neglect.

31

• • • • •

Cortexed but lone in spaced position,
Contain the hub of all audition,
Lesion me and eat word salad,
And vivid memories now pallid.

Hint #1:

These paired regions do not touch.

Hint #2:

Affected by the inferior M2 branch.

Temporal Lobe

The temporal lobe is the only lobe of the cerebrum not touching its paired lobe, thus *alone in spaced position to its paired lobe.* It *contains the primary auditory cortex* as well as Wernicke's area, which if *lesioned will produce a fluent aphasia* with speech containing nonsensical words. It also contains the hippocampus that aids in memory, thus *if lesioned, can produce memory impairment.*

The temporal lobe is broadly divided into a mesial compartment consisting of the hippocampus, amygdala, entorhinal, piriform, perirhinal, and parahippocampal cortex; a lateral compartment, consisting of the superior, middle, and inferior temporal gyri; and a basal compartment, consisting of lingual (medial occipito-temporal) and fusiform (lateral occipito-temporal) gyri. The temporal lobe is separated from the frontal and parietal lobes by the sylvian fissure. The opercular part of the temporal lobe contains the primary auditory cortex in the Heschl's gyri, which are a group of transverse gyri.

The mesial temporal lobe mostly consists of allo and mesocortex (less than six layers), whereas the basal and lateral temporal lobes are mostly made up of six-layered neocortex.

In the dominant hemisphere, in addition to the Wernicke's area (situated in the posterior superior and middle temporal gyri), there is a basal temporal language area as well, situated anteriorly within the fusiform gyrus. Posterior to this basal temporal language area is the visual language area in the fusiform gyrus. Lesion or electrical stimulation of this area produces pure alexia (without agraphia), which is the inability to read without aphasia.

32

• • • • •

Tent atop and rear confined,
Lesion left, now rightward blind,
As I contain a spur-shaped furrow,
And process what you see thorough.

Hint #1:

The Latin word for spur is *calcarine*.

Hint #2:

Say contralateral homonymous hemianopia five times fast.

Occipital Lobe

The occipital lobe *sits on top of the tentorium cerebelli* and is the *most posteriorly located lobe* of the brain. Because it contains the *primary visual cortex*, a lesion to one side *causes contralateral homonymous hemianopia*, thus lesion in the left results in right side blindness in both eyes. It also contains the *calcarine (Latin: spur) sulcus (Latin: furrow)*.

The visual information carried by the optic nerves is organized in a retinotopic fashion, and this organization is maintained all the way up to the primary visual cortex. After decussating in the optic chiasm, the optic tract then synapses in the lateral geniculate nucleus of the thalamus, from whence the optic radiations arise.

The optic radiations carrying information from the superior quadrants (inferior quadrants of visual field) of the corresponding halves of the two retinas end up in the superior bank of the calcarine fissure and the optic radiations from the inferior quadrants (superior quadrants of visual field) end up in the inferior bank of the calcarine fissure. Hence, a focal lesion of the superior bank of the calcarine fissure (cuneus) will lead to a contralateral inferior quadrantanopia, and a similar lesion of the inferior bank of the calcarine fissure (lingual gyrus) will lead to a contralateral superior quadrantanopia.

The primary visual cortex or area V1 (Brodmann area 17) is in the immediate pericalcarine cortex, and this is where the initial basic visual processing occurs. It is surrounded by Brodmann areas 18 and 19 (V2 and V3). There are many other association visual areas (V4–V8) located in the temporal and parietal lobes that perform very specialized functions in the visual pathway.

33

• • • • •

I run from mid to lat, right under your hat.
I split once and then again, and you'll know it's me just when,
Half your hemisphere, dark shaded will appear,
On the CT viewed, if proximal occlude.

Hint #1:

What did you ASPECTS?

Hint #2:

A friend of Mr. Willis.

Middle Cerebral Artery

The middle cerebral artery branches from the internal carotid artery and travels *laterally*, branching several times to perfuse the frontal, parietal, and temporal lobes unilaterally. If occluded proximally, it can result in the infarction of *~half of the cerebral hemisphere*, which is seen as a *hypodense region* on a non-contrasted CT scan.

As the middle cerebral artery originates from the terminal internal carotid artery bifurcation, it travels laterally toward the insula. This is called the M1 segment. The lenticulostriate arteries that supply the basal ganglia and internal capsule originate from the M1 segment. It also gives off the orbitofrontal and anterior temporal branches.

At the limen insulae, the artery (M2 segment) divides into a superior and inferior division supplying the majority of the cerebral hemisphere above and below the sylvian fissure. The M2 segment is also called the insular segment. The M3 segment (part of the superior division) courses laterally toward the sylvian fissure and is known as the opercular segment. The distal branches are referred to as the M4 segment.

The typical presentation of an MCA territory infarction is contralateral arm and leg weakness with contralateral facial droop with gaze deviation toward the side of the lesion. If the infarction is in the dominant hemisphere, patients often have global aphasia at presentation as well (both Broca's area and Wernicke's area are within the MCA vascular territory).

This is shown in Figures A.2 and A.3.

34

• • • • •

As it runs from Post to Ant,
Branched from busy three-way split,
An infarct will make leg weak,
And personality takes a hit.

Hint #1:

Think homunculus.

Hint #2:

Worst case scenario, akinetic mutism.

Anterior Cerebral Artery

The anterior cerebral artery (ACA) travels within the interhemispheric fissure, posterior to anterior from the bifurcation of the terminal internal carotid artery (ICA). The first segment of the ACA is short and extends from its origin at the bifurcation of terminal ICA to its junction with the anterior communicating artery (ACOM). From there, it almost runs vertically, rostral to the corpus callosum till its genu. It gives off a branch at this point called the callosomarginal artery. From here on, it continues as pericallosal artery running immediately above the corpus callosum within the callosal sulcus.

The recurrent artery of Heubner is a branch that comes off the second segment of the ACA distal to its junction with ACOM. It supplies the head of the caudate nucleus and the superior segment of the anterior limb of the internal capsule.

The typical presentation of a unilateral ACA territory infarction is contralateral leg weakness (the leg is represented in the primary motor cortex on the mesial surface of the frontal lobe). If the infarction is in the dominant hemisphere, patients can also have transcortical motor aphasia (impaired fluency, intact comprehension, and intact repetition).

Bilateral ACA occlusion can present with akinetic mutism. It is a state of behavioral arrest with severe impairment of spontaneous self-initiated movements and speech.

This is shown in Figures A.2 and A.3.

35

• • • • •

Vessel for flow that comes from post,
If infarcted, you'll lose the most,
Only eye motion will remain,
I sit midline, midpart, midbrain.

Hint #1:

Lock in to the right answer.

Hint #2:

Mr. Willis's spine.

Basilar Artery

The basilar artery sits *midline, on the ventral surface of the pons*, and is formed by a combination of two vertebral arteries. As it reaches the rostral midbrain, it splits terminally into two posterior cerebral arteries. Small perforating arteries coming off the basilar artery supply blood to the pons.

Before it terminally splits into the posterior cerebral arteries, the basilar artery gives off the anterior inferior cerebellar artery and the superior cerebellar artery.

A complete occlusion of the basilar artery will lead to a large infarction of the pons. These patients are often comatose in the acute period. If they survive the acute period and recover from the coma, they can end up in a state called "locked-in syndrome," characterized by quadriplegia, horizontal gaze palsy in both directions with relatively preserved vertical gaze.

Another important clinical syndrome known as top of the basilar syndrome is characterized by thromboembolism of the top of the basilar artery, leading to multifocal infarctions in bilateral thalami, midbrain, and bilateral occipito-temporal regions, producing a variety of clinical symptoms and signs including oculomotor disturbances, visual field deficits, and behavioral abnormalities, including impaired arousal.

This is shown in Figure A.3.

36

• • • • •

Ten grouped pairs of radicles,
From spinalis to last thread,
Oft pain when discs at risk, pinch brisk,
Opp side of the horse head.

Hint #1:

"Filum Terminale is Latin for terminal thread."

Hint #2:

Do not be a horse's ass, but close.

Cauda Equina

The cauda equina contains *10 pairs of nerve roots* (aka radicles, from Latin *radix* or root) that travel inferiorly from the *spinal cord* (spinalis) to the filum terminale (Latin meaning "terminal thread"). Cauda equina syndrome is a clinical syndrome of polyradiculopathy where the nerve roots are *often pinched by herniated discs*. Cauda equina derives its name from Latin due to its resemblance to a *"horse tail"* (at the opposite end of a horse from its head).

All the nerve roots below L1 (L2–L5 and sacral roots) constitute the cauda equina. The clinical symptoms are related to the compression of these roots. If there is predominantly sacral root involvement S2–S5, the chief symptoms are sensory and autonomic. Motor symptoms become prominent because of the involvement of lumbar roots.

The symptoms of cauda equina syndrome include low back pain and leg pain. Saddle anesthesia (buttocks, perineum, and posterior thighs) is present in >90% of patients. This area is represented by the coccygeal and S3–S5 nerve dermatomes. Urinary and bowel dysfunction are often present as well. The urinary incontinence is mainly an overflow incontinence because of a distended flaccid bladder. The normal voiding reflex depends upon the activation of the detrusor muscle by the parasympathetic fibers originating at the S2–S4 level.

Lower extremity weakness of any pattern can be present (symmetric or asymmetric) and reflexes may be diminished. The diagnostic test of choice is MRI of the lumbosacral spine.

The most common cause of cauda equine syndrome is lumbar disc herniation. Urgent neurosurgical consultation is required. Early surgical intervention is thought to improve the neurological outcome although studies have been equivocal. The most important prognostic factor is the degree of the neurological dysfunction prior to surgery.

37

• • • • •

I do not contain any mass, myself,
A valley twix two great mounds,
I separate the doers and feelers,
Located on both half-rounds.

Hint #1:

Paired and almost mirror-imaged structures.

Hint #2:

Never touch, but come close.

Central Sulcus

The central sulcus is a furrow *separating the precentral gyrus from the postcentral gyrus (two mounds)*. Each hemisphere (half-rounds) contains a central sulcus.

The central sulcus is an important anatomical landmark separating the frontal lobe (doers) from the parietal lobe (feelers).

The anterior bank of the central sulcus mainly consists of Brodmann area 4 (primary motor cortex). The fundus consists of Brodmann area 3a, which is the transition between the primary motor cortex and the primary sensory cortex. The posterior bank of the central sulcus consists of Brodmann area 3b, which is part of the primary sensory cortex. Overall, the precentral gyrus mainly consists of Brodmann areas 4 and 6, which together constitute the primary motor cortex and the postcentral gyrus consisting of Brodmann areas 3, 1, and 2 (primary sensory cortex).

From a cytoarchitectural standpoint, Brodmann area 4 is a granular cortex. Area 3a starts showing evidence of granular layer IV, and area 3b onwards is proper granular cortex with a thick layer IV.

Although the central sulcus separates the two gyri, there are deep-buried, interlocking gyri known as annectant gyri, and this phenomenon is referred to as plis de passage. In rare cases, these submerged gyri rise to the surface and interrupt the central sulcus, the most famous example being plis de passage frontal moyen, first described by Broca. This is located in the middle central sulcus and is commonly referred to as the "hand knob."

This is shown in Figure A.1.

38

• • • • •

Midline, sitting inside a saddle,
Just below the crossing sees,
Connected above, by one small stalk,
With hormones making you and me.

Hint #1:

This saddle was made by a Turk.

Hint #2:

"Sees" is not a typo.

Pituitary Gland

The pituitary gland is *positioned midline inside the sella turcica* (Latin meaning "Turkish saddle"). It is positioned just *below the optic chiasm* and connected to the rest of the brain by the *pituitary stalk.* It *functions as an endocrine gland,* producing many hormones such as adrenocorticotrophic hormone, follicle-stimulating hormone, luteinizing hormone, growth hormone, prolactin, thyroid-stimulating hormone, oxytocin, and antidiuretic hormone, which have an enormous importance in multiple body functions.

The sella turcica is formed by the sphenoid bone surrounded by anterior and posterior clinoid processes. Underneath the sella is the sphenoid sinus. The layer of dura covering the pituitary fossa is known as diaphragma sella. The cavernous sinus with its important structures (internal carotid artery, cranial nerves III, IV, V1, V2, and VI) is located on both sides of the pituitary fossa.

Because of its proximity to the optic chiasm and optic nerves, visual field defects are important clinical exam findings in patients with pituitary adenomas. For example, bitemporal hemianopsia caused by damage to the crossing nasal fibers at the optic chiasm. Because this is a heteronymous field defect, the two adjacent residual nasal fields do not overlap and can slide, thus leading to a nonparetic form of diplopia known as retinal hemifield slide phenomenon. In addition, pituitary tumors can produce multiple cranial nerve pathologies by invading the surrounding cavernous sinus.

This is shown in Figure A.1.

39

• • • • •

Split from the base, and wrapped around tres,
Feed mid, then back I will go,
You can write but not read it, if flow is impeded,
If on left then right vision will not show.

Hint #1:

Close friends with CN3.

Hint #2:

Think alexia without agraphia.

87

Posterior Cerebral Artery

The posterior cerebral artery (PCA) is the terminal branch of the basilar artery (and thus *split from the "base"*) and *wraps around the oculomotor nerve* (CN III). It *perfuses the midbrain,* then *travels back to the occipital lobes.*

The PCA has four or five segments. The P1 segment extends from the origin at the basilar artery to the junction with the posterior communicating artery. P2 wraps around the crus cerebri in the pericrural cistern and transits through the ambient cistern. The P3 segment travels in the quadrigeminal cistern. P4 is the cortical segment that nourishes the occipital lobe. The main branch is the calcarine artery traveling in the calcarine fissure. The terminal branches are referred to as the P5 segment.

Occlusion of the left PCA will lead to infarction of the left occipital lobe, which, in turn, will lead to a right homonymous hemianopia (contralateral field cut). Since the PCA also supplies the splenium of the corpus callosum, this region can also get infarcted. A combination of left occipital lobe and splenium infarction can produce a peculiar syndrome of alexia without agraphia. These patients aren't aphasic and can write as well. They have an isolated deficit of reading. This occurs because the visual information from the intact right occipital lobe (seeing the left hemifield) cannot reach the language-dominant left hemisphere via the splenium of corpus callosum. Some of these patients also have associated color anomia, which localizes to the lingual and fusiform gyri.

This is shown in Figure A.3.

40

· · · · ·

The strobile of nerves at the joining of curves,
You'll see me end right around two,
Dysfunction and numbness and sexual glumness,
Tap too high and I will feel you.

Hint #1:

The strobile comes from the Latin word *strobilus*, meaning "cone."

Hint #2:

Think saddle anesthesia/paresthesia.

Conus Medullaris

OK! OK! I know this one was tough. The conus medullaris is the cone located at the base of the spinal cord. *Strobile* is the Latin word for cone, and it is located around the meeting of the kyphotic (thoracic spine) curve and the lordotic (lumbar spine) curve. It ends around L1–L2.

The human spinal cord has 8 cervical segments, 12 thoracic, 5 lumbar, 5 sacral, and 1 coccygeal segment. During embryological development, the vertebral column continues to grow even after the spinal cord has fully formed. Hence, the 5 sacral and 1 coccygeal segments of the spinal cord end at the level of L1 or L2 vertebra and constitute the conus medullaris.

Conus medullaris syndrome is clinically similar to cauda equina syndrome, except for the fact that conus medullaris syndrome is a form of myelopathy since the pathology affects the spinal cord, whereas the cauda equina syndrome is a polyradiculopathy.

The symptoms include *perineal numbness, bowel/bladder dysfunction, and sexual dysfunction*. The conus medullaris can be injured if a lumbar puncture is performed at a level above the L1–L2 vertebra.

41

• • • • •

The tube with the tracks, that goes down our backs,
Past levels of about twenty-one,
An electrical nexus that helps with reflexes,
Cut me and your walking is done.

Hint #1:

All Cs, all Ts, some Ls.

Hint #2:

Some say contains a butterfly.

Spinal Cord

The spinal cord is encased in the vertebral column and contains *ascending and descending white matter tracts* from the brain. It also contains alpha motor neurons, which give rise to nerve fibers controlling muscles.

The spinal cord consists of 8 cervical, 12 thoracic, 5 lumbar, 5 sacral, and 1 coccygeal segment. At the level of the cervical cord, each nerve root exits above the level of the corresponding vertebra. For example, the C6 nerve root exits above the C6 vertebra and below the C5 vertebra. At the level of the thoracic and lumbar cord, each root exits below the level of the corresponding vertebra.

The spinal cord is not as long as the vertebral column and actually ends at the L1 or L2 vertebral level. The sacral and coccygeal segments form the conus medullaris. The nerve fibers of the lumbar and sacral segments constitute the cauda equina.

As opposed to the brain, in the spinal cord, gray matter is surrounded by white matter. The gray matter is organized into a ventral horn and a dorsal horn with an intermediate zone. The dorsal horn receives sensory fibers from the dorsal ganglia located outside the spinal cord. The ventral horn contains the alpha motor neurons, and the intermediate zone contains the sympathetic neurons.

The lateral corticospinal tract is located in the lateral funiculus, whereas the more primitive motor pathways like the vestibulospinal tract, tectospinal tract, and reticulospinal tract are located in the ventral funiculus. The dorsal column sensory pathways are located in the posterior funiculus, whereas the spinothalamic tract is located in the lateral funiculus.

Diseases affecting the spinal cord manifest as motor symptoms (weakness), sensory disturbance, and urinary and/or bowel incontinence. On examination, these patients often have spasticity, hyperreflexia with or without clonus, and upgoing toes. This constellation of examination findings is referred to as myelopathy.

This is shown in Figure A.1.

42

• • • • •

Two doers, not feelers, from your toe to your tongue,
Controlling most muscles, but not intestines or lung,
Apart by a trench, from feel center renown,
In the back of the front, where millions go down.

Hint #1:

Intestines and lungs are primarily under involuntary muscle control.

Hint #2:

Think insensitive homunculus.

Primary Motor Cortex/Precentral Gyrus

The primary motor cortex (M1) is located in the precentral gyrus in the frontal lobe. It is separated from the primary sensory cortex (S1) by the central sulcus. It is a six-layered agranular cortex. It consists of Brodmann areas 4 and 6. Area 4 is mainly located in the anterior bank of the central sulcus, whereas area 6 is located on the crown of the precentral gyrus.

The primary motor cortex contains a somatotopic representation of the contralateral body. This is best represented by the so-called Penfield homunculus, which was first described by the neurosurgeon Wilder Penfield at the Montreal Neurolgical Institute at McGill University using intraoperative direct electrical cortical stimulation of epilepsy patients. This homunculus consists of disproportionately large representations of face and hand on the lateral convexity with relatively smaller representations of leg and foot on the mesial surface. This map is generally represented in a linear contiguous fashion, although recent studies have shown that the face, hand, and foot areas are organized as concentric effector regions, interrupted by inter-effector regions.

The neurons in the primary motor cortex give rise to the corticospinal tract, which descends via the brainstem, decussates in the medulla, and then synapses on the alpha motor neurons in the spinal cord to execute planned movements.

43

• • • • •

I integrate info from millions,
The third stop on two ascending path,
Disproportioned hands and face,
How you'll feel your nice warm or cold bath.

Hint #1:

The highest point in the sensory pathways.

Hint #2:

Think sensitive homunculus.

Postcentral Gyrus/Primary Sensory Cortex

The primary sensory cortex (S1) is the last stop (*numerically the third stop*) and place of information integration/interpretation of the *millions of fibers in the ascending dorsal column-medial lemniscal pathway.*

S1 is primarily a six-layered granular cortex. It consists of Brodmann areas 3b, 1, and 2. Area 3b is present in the posterior bank of the central sulcus, whereas areas 1 and 2 are present in the crown of the postcentral gyrus. Area 3a is present in the fundus of the central sulcus and is a transitional zone between the primary motor cortex (Brodmann area 4) and S1.

The granularity of S1 is mainly because of a thick layer IV. This layer contains neurons that receive afferents from the thalamus.

Much like the motor homunculus, the primary sensory cortex has its own homunculus, with the areas that are most superomedial (areas within the longitudinal fissure) integrating *sensation from your genitals and toes*, and the region that is most inferolateral integrating *sensation from the face and tongue*. Similar to the motor homunculus, the face and hand regions have disproportionately large representations compared to the rest of the body.

44

• • • • •

Two beans, side-by-side, adhesed by a mass,
And also called the bridal couch,
The lateral walls of the third,
Stop two when nerves come up, go ouch.

Hint #1:

Bean-shaped, but much bigger.

Hint #2:

Important stop for the ascending spinothalamic pathway.

Thalamus

The thalamus (the Greek word meaning *bridal couch or marriage chamber*) is a paired structure that many refer to as "bean-shaped." The pairing lie *lateral to one another* and are joined by the interthalamic *adhesion,* also called the intermediate *mass.* The thalamus composes the *lateral walls of the third ventricle* and is the location of the *second synapse in the ascending sensory pathways.*

It is a derivative of diencephalon along with the hypothalamus, subthalamus, epithalamus, and the optic nerves.

The thalamus is a collection of nuclei. Most of the nuclei are called relay nuclei. There are three main groups – medial, lateral, and anterior. These are separated by a Y-shaped structure known as the internal medullary lamina, which contains the intralaminar nuclei. In addition, the thalamus is covered by a thin layer known as the reticular nucleus.

The somatosensory pathways synapse in the ventropostero-medial (VPM) and VPL nuclei. The VPM relays sensory information from the face, and the VPL relays sensory information from the body.

The auditory information is relayed via the medial geniculate nucleus, whereas the visual information is relayed via the lateral geniculate nucleus.

The reticular nucleus plays a critical role in state-dependent regulation of the activity of other thalamic nuclei.

This is shown in Figure A.2.

45

• • • • •

Capped above by cingulates,
A midline, C-shaped, sideway pass,
Four parts or five, from beak to patch,
White matter concentrated mass.

Hint #1:

The Greek word for patch or bandage/plaster is *splenium*.

Hint #2:

Communication between the two cerebral hemispheres.

Corpus Callosum

The corpus callosum is a *midline, C-shaped structure* that sits *below the cingulate gyrus*. It functions as a *byway for communication between the two cerebral hemispheres*. It *contains four main parts*, namely rostrum (named in Latin for its "beak-like shape"), genu, trunk/body, and splenium (Greek for patch or bandage/plaster). Some anatomists, however, also include the isthmus of the corpus callosum, which is the narrowed part between the body and the splenium.

The corpus callosum is the *largest collection of white matter tracts* anywhere in the brain.

Since corpus callosum connects areas of cortex between the two hemispheres, lesions affecting the corpus callosum can present with unusual signs and symptoms referred to as disconnection syndromes. A classic example is left-hand apraxia caused by an anterior corpus callosum lesion, which disconnects the dominant language areas of the left hemisphere from the premotor cortex in the right hemisphere, thus interfering with motor programming of the left hand. Other examples include alexia without agraphia and alien limb syndrome.

Isolated lesions of the corpus callosum are rare. Some of the differentials for isolated cytotoxic lesions of corpus callosum (CLOCC) include sudden cessation of antiseizure medications, metabolic abnormalities (hypo- or hypernatremia, hyperammonemia, hypoglycemia), toxic causes (Marchiafava–Bignami disease, Wilson disease), infarcts, tumors, and so on.

This is shown in Figure A.1.

46

• • • • •

In four of the five, made by the plexus,
Tested when the elbow flexes,
With sensory to top of shoulder,
Specifically not within ulnar.

Hint #1:

Common starting place for the neurological reflex exam.

Hint #2:

Erb's Palsy.

C5 Nerve Root

The C5 nerve root is contained within *four of the five peripheral nerves that come from the brachial plexus* (musculocutaneous, median, radial, and axillary, *not ulnar*). Its function is commonly tested during neurological exam by tapping the biceps tendon with a reflex hammer, which *creates reflexive elbow flexion*. The C5 dermatome is positioned on the *superior and lateral shoulder*.

The C5 root joins the C6 root to form the upper trunk of the brachial plexus. The anterior division of the upper trunk joins the anterior division of the middle trunk (C7) to form the lateral cord, which contributes to the formation of the median nerve and continues as the musculocutaneous nerve. The posterior division joins the other posterior divisions to form the posterior cord, which gives rise to the radial nerve.

Important muscles containing the C5 root include rhomboids, supraspinatus, infraspinatus, and deltoid.

Flexion of the elbow in a supinated position is mainly a function of biceps brachii, which is innervated by the musculocutaneous nerve (C5 root), whereas elbow flexion in the semi-pronated position is a function of brachioradialis, which is innervated by the radial nerve (C5 root). This simple test can distinguish between elbow flexion weakness caused by musculocutaneous neuropathy or lateral cord lesion versus radial neuropathy or posterior cord lesion. A lesion of the upper trunk of the brachial plexus or the C5 root will cause weakness of both muscles though.

A lesion of the upper trunk of the brachial plexus can be distinguished from a lesion of the C5 nerve root by testing the strength of the rhomboids, which are innervated by the dorsal scapular nerve (C5), which comes off the root prior to the plexus.

47

• • • • •

One of five within the rete,
And using me, four nerves complete,
Some say pointer but all say thumb,
Change in jerk if this root goes bum.

Hint #1:

Think dermatomes.

Hint #2:

Some test in reflex exam.

C6 Nerve Root

The C6 nerve root is one of the roots *contributing to the brachial plexus* (also known as the *rete plexus*). And just like the C5 nerve root, its fibers are *contained within four of the five peripheral branches* of the brachial plexus.

Most resources state that the index finger is included in the C6 sensory dermatome; however, *all sources agree that the thumb is included in it.*

The C6 root joins the C5 root to form the upper trunk of the brachial plexus. The anterior division of the upper trunk joins the anterior division of the middle trunk (C7) to form the lateral cord, which contributes to the formation of the median nerve and continues as the musculocutaneous nerve. The posterior division joins the other posterior divisions to form the posterior cord, which gives rise to the radial nerve.

Important C6 innervated muscles include brachioradialis and extensor carpi radialis.

Its function is often tested in the neurological exam by tapping the brachioradialis tendon, which produces a slight elbow flexion and supination known as the *brachioradialis reflex (aka supinator reflex).*

A clinical sign known as the inverted radial reflex consists of finger flexion when tapping the brachioradialis tendon with absent elbow flexion as would be normally expected. This is a myelopathic sign that localizes to the C5–C6 spinal cord level. The mechanism of finger flexion is thought to be because of the pathological spread of the reflex to C8 level, while the absence of elbow flexion is because of the lesion being at the level of C5–C6.

48

• • • • •

I'm the third down of five radix,
Sense middle finger, back of arm,
From the start becomes the middle,
Tap back of elbow, I'm alarmed.

Hint #1:

Think dermatomes.

Hint #2:

The only one in this trunk.

C7 Nerve Root

The C7 nerve root is the third root from the top in the brachial plexus. The word root is derived from Latin *radix*. The C7 dermatome provides sensation to the middle finger and a thin strip of sensation on the back of the arm.

The C7 root by itself constitutes the middle trunk of the brachial plexus. Its anterior division joins the anterior division of the upper trunk to form the lateral cord. Its posterior division joins the posterior divisions of upper and lower trunks to form the posterior cord, which gives rise to the axillary and radial nerves.

Important C7 innervated muscles include triceps and all wrist and digit extensors in the posterior compartment of the forearm. All these muscles are innervated by the radial nerve, which is the terminal extension of the posterior cord.

The function of the C7 root is tested via the triceps reflex by tapping the back of the elbow with a reflex hammer.

49

• • • • •

From the bottom of my region,
I join the top of one below,
Lower trunk, part of three nerves,
Sense pinky, and flex fingers, go.

Hint #1:

"region" = spinal cord region.

Hint #2:

Fibers become part of the lower trunk and medial cord.

C8 Nerve Root

The C8 nerve root is the lowest root in the cervical spine. After exiting the cord, it joins the T1 nerve root below it (which is the first root of the thoracic spine) to form the lower trunk of the brachial plexus.

The posterior division of the lower trunk joins the posterior divisions of the upper and middle trunks to form the posterior cord. The C8–T1 fibers in the radial nerve (which comes off the posterior cord) come from the posterior division of the lower trunk.

The anterior division of the lower trunk continues as the medial cord. The medial cord joins the lateral cord to form the median nerve. The C8–T1 fibers in the median nerve come from the medial cord. The rest of the medial cord continues as the ulnar nerve.

This is how the C8 root provides fibers to three major peripheral nerves – ulnar, radial, and median nerves. The C8 dermatome provides sensation to the little finger and the ring finger.

Most of the intrinsic hand muscles are innervated by the ulnar nerve (C8, T1). A C8 radiculopathy can easily mimic an ulnar neuropathy. On clinical exam, ulnar neuropathy can be distinguished from C8 radiculopathy by testing the abductor pollicis brevis (APB) muscle strength and extensor indicis proprius muscle strength. The APB is innervated by the median nerve but has C8 root value. The extensor indicis proprius also has C8 root value, but is innervated by the posterior interosseous branch of the radial nerve.

50

· · · · ·

I'm the top of second region,
Lowest one of five in bunch,
Add and abduct all your fingers,
A numb elbow if this root crunch.

Hint #1:

Consider the brachial plexus.

Hint #2:

Not primarily tested in the reflex exam.

T1 Nerve Root

The T1 nerve root is the most superior root in the second region of the spine, the thoracic region. The T1 root joins the C8 root to form the lower trunk of the brachial plexus.

The posterior division of the lower trunk joins the posterior divisions of the upper and middle trunks to form the posterior cord, which terminates as the radial nerve.

The anterior division of the lower trunk forms the medial cord. The medial cord joins the lateral cord to form the median nerve and then continues as the ulnar nerve.

The T1 innervated muscles in the hand supplied by the median nerve include the first two lumbricals, opponens pollicis and abductor pollicis brevis, and the superficial head of the flexor pollicis brevis.

The T1 innervated muscles in the hand supplied by the ulnar nerve include third and fourth lumbricals, palmar and dorsal interossei, the deep head of flexor pollicis brevis, adductor pollicis, and abductor digiti minimi.

51

• • • • •

A spatial mapping true asset,
Named from horse that's in the sea,
Like H.M. you might forget,
And lose your way if injure me.

Hint #1:

Significantly affected in Alzheimer's disease.

Hint #2:

Resembles a seahorse.

Hippocampus

The hippocampus is located in the mesial temporal lobe and is an incredibly important hub for memory formation, especially in the encoding of new memories.

The name "hippocampus" is derived from the Greek *hippos* meaning "horse" and *kampos* meaning "sea monster," due to its shape.

The hippocampal formation located within the mesial temporal lobe has an S-shape in the coronal plane and consists of the dentate gyrus, cornu ammonis (hippocampus proper), and the subiculum. The other important structure in this region is the parahippocampal gyrus, which contains the entorhinal cortex. The cornu ammonis is divided into three sectors – CA1 to 3. The dentate gyrus is also referred to as CA4.

The two main input pathways for the hippocampus are the perforant pathway and the alvear pathway. The perforant fibers arise in the pyramidal neurons of the entorhinal cortex, cross the subiculum and the hippocampal sulcus, and synapse directly on the granule cells of the dentate gyrus. These cells, in turn, give rise to the mossy fibers that synapse onto the dendrites of the CA3 neurons. The axons arising out of these neurons constitute the major output pathway of the hippocampus leaving the hippocampal formation via the fornix. The CA3 neurons also give collaterals that synapse on CA1 neurons, known as Schaffer collaterals. This intrinsic circuitry is known to exhibit the property of long-term potentiation (LTP), which is thought to play a role in strengthening synaptic plasticity and hence memory.

Patient H.M. was operated upon by Dr. Scoville at Hartford hospital for medically refractory epilepsy. He underwent resection of bilateral hippocampi and ended up with severe anterograde amnesia. This case really highlighted the role of the hippocampus in memory function.

Research has shown that the hippocampus plays a very important role in spatial navigation as well. This is based on the discovery of place cells in the hippocampus and grid cells in the medial entorhinal cortex.

52

• • • • •

A Greek "almond," deep temporal,
Governs emotional reaction,
If both impaired, you may get scared,
Kluver–Bucy, and lewd action.

Hint #1:

Lesions can produce oral fixation and hypersexuality.

Hint #2:

A key component of the limbic system.

Amygdala

The amygdala derives its name from the Greek word *amygdale* for "almond," given to it because of its almond-like shape. It is located in the mesial temporal lobe.

The amygdala is a collection of nuclei and, embryologically, has the same origin as the basal ganglia (strio-amygdaloid complex). The amygdala is intricately connected to the hippocampus, the rest of the temporal lobe, insula, frontal lobe, and the hypothalamus. Thus, it plays a very important role in emotion and autonomic control.

The three main amygdaloid nuclei are corticomedial, basolateral, and central. The central nucleus is mainly concerned with autonomic functions and is connected with the hypothalamus and the brainstem. The corticomedial nucleus is mainly involved in processing olfactory information. The basolateral nucleus is the largest in humans and has widespread connections to many areas of the cortex, including strong connections with the hippocampus.

The two main efferent pathways of the amygdala include the stria terminalis and the ventral amygdalofugal pathway. The stria terminalis connects the amygdala to the hypothalamus and the septal nuclei. The ventral amygdalofugal pathway connects it to the hypothalamus, brainstem, limbic striatum, and nucleus basalis of Meynert. Other efferent pathways include the anterior commissure, connecting it to the contralateral amygdala, and the uncinate fasciculus, connecting it to the orbitofrontal cortex.

Both stria terminalis and the ventral amygdalofugal pathway have similar targets, but the stria terminalis is a longer C-shaped tract, traveling medial to the caudate nucleus.

Isolated unilateral lesions of amygdala are seldom symptomatic, but bilateral lesions are associated with Kluver–Bucy syndrome, which is characterized by oral fixation, hyperorality, timid behavior, and hypersexuality.

In addition, the amygdala is very commonly the chief focus of epileptogenicity in patients with temporal lobe epilepsy with or without hippocampal involvement. Seizures arising from the amygdala are very commonly associated with oroalimentary automatisms with loss of awareness. The amygdala is also implicated in pathogenesis of primary central ictal apnea in temporal lobe epilepsy.

53

• • • • •

I am an *egg-shaped* hole,
In a Greek-named *wedge-shaped* bone,
Through me there passes just one thing,
Fifth nerve, third branch, alone.

Hint #1:

Holds V3.

Hint #2:

The Greek word *sphenoeides* means "wedge-shaped" and is the origin of the name given to the sphenoid bone.

Foramen Ovale

I know, I know! This one is super tough, solely based on the need to look up Latin and Greek origins to make sense of it. The foramen ovale gets its name from the early Latin word *ovum* meaning egg, which would later become *ovalis* and then the word "oval," which is an egg-shaped circle. It is positioned within the sphenoid bone which means "wedge-shaped," *sphenoeides* in Greek. Only V3, the third branch of the trigeminal nerve, passes through it.

The V3 or the mandibular division of the trigeminal nerve carries both sensory and motor fibers. The sensory fibers innervate the chin and the lower jaw area, and the motor fibers innervate the muscles of mastication such as the masseter.

Numb chin syndrome is a pure sensory neuropathy of the mental nerve, which is a branch of the V3 division of the trigeminal nerve. It is characterized by isolated numbness over the chin and part of the lower lip. The most common etiology is underlying malignancy with metastatic lesion to either the mandible affecting the mental nerve or the skull base near the foramen ovale affecting the V3 division as it exits the skull.

54

• • • • •

A black substance near the third nerve,
I make dopa, help you move,
Lose my cells brings shuffled feet,
Replete dopa, you'll improve.

Substantia Nigra

The substantia nigra gets its name from a Latin word meaning "black substance." This is because when viewed as an unstained specimen, it appears as a darkened streak. Located in the midbrain, it lies lateral to the oculomotor nerve as it leaves the brainstem. It is further made up of two parts – substantia nigra pars compacta (SNpc) and substantia nigra pars reticularis (SNpr).

The SNpr works as the main output organ of the basal ganglia along with globus pallidus interna. The key function of the SNpc is to produce dopamine and project it to the striatum. It plays a major role in neuromodulation, helping with motor control and reward function.

The substantia nigra is an important and early site of pathological alpha synuclein accumulation in idiopathic Parkinson's disease. This leads to the characteristic motor symptoms of bradykinesia and tremor. Key to treating this condition is the repletion of dopamine, which improves these symptoms.

55

• • • • •

A branch'ed sanguine conduit,
My friend Max is where I seed,
Up I go past the Greek wing,
If I'm ruptured, lens-shaped bleed.

Hint #1:

The word "sanguine" is derived from the Latin *sanguis*, meaning "blood."

Hint #2:

The Greek word *pteron*, means "wing," and the origin of the word pterion.

Middle Meningeal Artery

The middle meningeal artery (artery being a carrier of sanguine, which is derived from the Latin *sanguis*, meaning "blood") is an artery with many branches, which perfuses two-thirds of the cranial dura. It is a branch of the maxillary artery (a branch of the external carotid artery), which travels superiorly, enters the middle cranial fossa via the foramen spinosum, and on its way, passes the pterion, which derives its name from the Greek word *pteron*, meaning "wing."

The pterion is a common location for skull fracture and subsequent rupture of the middle meningeal artery, causing an epidural hematoma. On CT scan, epidural hematomas are known for their characteristic lens-shaped appearance as blood collects between the dura and the skull.

Middle meningeal artery embolization is a neurosurgical procedure used for patients with recurrent chronic subdural hematoma. Before performing the embolization procedure, it is essential to establish the presence of variant anatomy, that is, the origin of the ophthalmic artery from the middle meningeal artery, rather than from the internal carotid artery directly. The inadvertent occlusion of the ophthalmic artery can cause global ocular ischemia, resulting in irreversible blindness.

56

• • • • •

A hole through which three things do pass,
Most post- in the great wing,
One nerve, two vessels, with three diff' jobs,
To tough mother through this small ring.

Hint #1:

Follows the middle meningeal artery.

Hint #2:

Think V3.

Foramen Spinosum

The foramen spinosum is the most posterior passage/hole in the greater wing of the sphenoid bone, posterolateral to the foramen ovale. The three structures that traverse this foramen are the middle meningeal artery, middle meningeal vein, and a meningeal branch of the V3 nerve (a mandibular division of the trigeminal nerve).

This nerve innervates the dura of the middle cranial fossa and is also known as nervus spinosus.

Each of the three structures performs separate functions, bring blood to the dura, draining blood from the dura, and innervating the dura, respectively. All of them, however, extend from the foramen spinosum to the dura mater (*tough mother* in Latin).

The middle meningeal artery, a branch of the maxillary artery, which is a branch of the external carotid artery, enters the skull via the foramen spinosum. This is often the culprit artery lacerated in cases of traumatic epidural hematomas.

57

• • • • •

Named for my quite round contour,
Just one thing that passes through,
In wasp or wedge is where I sit,
A bore that holds V2.

Hint #1:

Middle cranial fossa.

Hint #2:

Not an oblong circle.

Foramen Rotundum

The foramen rotundum is named after the Latin word *rotundum* meaning "round." The chief structure that traverses this foramen (or *bore*) is V2, the maxillary branch of the trigeminal nerve.

Foramen rotundum is located at the base of the greater wing of the sphenoid bone, which many believe derives its name from the Greek word *os sphecoidale*, meaning "bone resembling a wasp." Others believe that the word "sphenoid" originated from the word *sphen* meaning "wedge-shaped" in Greek. The foramen is inferior and medial to the superior orbital fissure.

A patient with isolated sensory symptoms in the V2 distribution would indicate the presence of a skull base tumor near the foramen rotundum. Dedicated imaging is warranted in such cases.

58

• • • • •

From the third of four sep'd regions,
Exit side and then descend,
Sensation to the medial leg,
Tap the knee, I help extend.

Hint #1:

Think spinal/vertebral regions.

Hint#2:

Think the stereotypical reflex hammer test.

131

L4 Nerve Root

The L4 nerve root originates from the lumbar spinal cord region, the third of the four regions of the spine from top to bottom. This nerve root is part of the cauda equina as the spinal cord ends at the L1–L2 vertebral level. It eventually exits through the intervertebral foramen between L4 and L5 vertebral bodies.

Some chief L4 innervated muscles include quadriceps femoris (knee extension), innervated by the femoral nerve, and thigh adductors, innervated by the obturator nerve. The function of the L4 root can be tested via the patellar reflex, by tapping the patellar tendon with a reflex hammer, which elicits a reflexive knee extension.

The L4 dermatome is mainly represented over the inner leg up to the medial malleolus.

Isolated L4 radiculopathy is rare, but it can mimic a femoral neuropathy. It can be clinically differentiated from femoral neuropathy by testing non-femoral L4 innervated muscles like the thigh adductors (obturator nerve). In addition, on electrodiagnostic testing, the sensory nerve action potentials are preserved in radiculopathies as opposed to mixed peripheral neuropathies.

59

• • • • •

The root most common hit,
By a posterolateral blow,
I give sensation to foot dorsum,
And extension to the great toe.

Hint #1:

Not posterior blow because of the posterior longitudinal ligament.

Hint #2:

Just slightly more common than S1.

L5 Nerve Root

The L5 nerve root is the lowest lumbar nerve root that exits between the L5 and S1 vertebrae. It is the most common root implicated in lower extremity radiculopathy. Because of the posterior longitudinal ligament, herniated discs most often protrude posterolaterally where they commonly compress nerve roots exiting through the intervertebral foramina.

The L5 dermatome is represented over the outer leg and the top of the foot. There are multiple muscles in the thigh and the leg that contain L5 fibers, and innervated by sciatic, common peroneal, and tibial nerves.

Clinically, patients with L5 radiculopathy often present with sensory symptoms over the dorsum of the foot and the lateral surface of the leg below the knee. In addition, they can have a foot drop because of weakness of foot and toe dorsiflexors.

A common peroneal neuropathy can have a very similar clinical presentation. The way to differentiate common peroneal neuropathy from L5 radiculopathy is to test muscles innervated by the tibial nerve with the same root value of L5. This is accomplished by testing foot inversion, which is a function of tibialis posterior, a tibial nerve muscle with L5 root value. In addition, the weakness of toe dorsiflexion is disproportionately worse compared to foot dorsiflexion weakness in L5 radiculopathy as opposed to common peroneal neuropathy.

Weakness of foot inversion rules out a common peroneal neuropathy, but it is not specific to L5 radiculopathy as it can be present with sciatic neuropathy or lumbosacral plexopathy as well. Sciatic neuropathy can be distinguished from L5 radiculopathy by testing the strength of gluteus medius (innervated by the superior gluteal nerve, L5 root) or tensor fasciae latae (also innervated by the superior gluteal nerve, L5 root).

One of the most important principles of electrodiagnostic testing in radiculopathies is that the sensory nerve action potential (SNAP) is almost always normal because the lesion in radiculopathy is proximal to the dorsal root sensory ganglion that sits outside the spinal cord. One exception to this important principle is L5 radiculopathy. Rarely, in L5 radiculopathy, the superficial peroneal SNAP can be abnormal. The reason is not clearly understood.

60

• • • • •

I'm the first within my region,
And second most common compressed,
If so, numbness, to the lateral foot,
And ankle reflex depressed.

Hint #1:

Think spinal/vertebral region.

Hint #2:

Can substitute pain for numbness.

S1 Nerve Root

The S1 nerve root is the first nerve root of the sacral spinal cord segment. It is the second most commonly implicated nerve root in radiculopathies caused by a herniated disc.

Like the L5 root, the S1 root innervates multiple thigh and leg muscles via the sciatic, common peroneal, and tibial nerves.

The S1 dermatome is mainly represented as a thin segment of the posterior aspect of the lower leg and the lateral and plantar aspects of the foot.

Patients with S1 radiculopathy often complain of radicular lower back pain radiating into lower leg and the bottom of the foot, consistent with S1 dermatome. Usually there are no major motor deficits because of the strong L5 representation of the foot dorsiflexors. However, slight weakness of the plantarflexors can be detected as the soleus is primarily of S1 and S2 myotome. In addition, the ankle jerk is usually diminished or absent.

During electrodiagnostic testing, a soleus H-reflex should be tested as well in suspected cases of S1 radiculopathy (electro-physiological correlate of the ankle jerk). And as mentioned previously, the Sensory nerve action potential (SNAP) will be normal in S1 radiculopathy as opposed to lumbosacral plexopathy or sciatic neuropathy.

61

• • • • •

I supply two-thirds the blood,
And rarely flow I'm bereft,
Though the other third supplied by two vessels,
If I clot, vibration's left.

Hint #1:

A branch of the verts.

Hint #2:

No pain/temperature or movement.

Anterior Spinal Artery

The anterior spinal artery is an artery that travels up the anterior side of the spinal cord, perfusing two-thirds of the spinal cord territory (notably the corticospinal and spinothalamic tracts). The other one-third of the perfusion territory of the spinal cord is supplied by the two posterior spinal arteries (notably the dorsal columns).

Spinal cord infarcts are relatively rare when compared to cerebral infarcts. Additionally, given that the dorsal columns are outside the perfusion territory of the anterior spinal artery, if an infarct were to occur, vibration and proprioception sense would be preserved.

Anterior spinal artery is formed rostrally at the level of medulla by the two vertebral arteries, and it descends on the ventral surface of the spinal cord. Throughout the length of the spinal cord, there are arteries arising from various other arteries, known as radiculomedullary or segmental medullary arteries that reinforce the blood supply to the spinal cord.

A very prominent radiculomedullary artery between T9 and T12 levels is the great radicular artery of Adamkiewicz. This artery can get damaged in cases of aortic dissection or during surgical procedures and can lead to spinal cord infarction at lower thoracic level.

62

• • • • •

The spawn to two large vessels,
In six and skull into,
Each side goes through seven diff' holes,
What post-circulation goes through.

Hint #1:

Start in the top of the thorax.

Hint #2:

Six of seven that is.

Vertebral Arteries

The vertebral arteries are the main source of the posterior cerebral circulation. Each vertebral artery originates as a branch of the subclavian artery in the thorax.

The vertebral artery has four segments. V1 is the segment from the origin to the point where the artery enters the transverse foramen of the C6 vertebra.

V2 refers to the segment of the artery from C6 to C2 vertebra as the artery traverses through each transverse foramen. V3 is the segment where the artery exits the transverse foramen of C2 vertebra, travels laterally over the transverse process and lateral to the C1–2 articulation, then ascends superiorly to go through the transverse foramen of C1.

V4 is the final segment as the artery enters the cranial cavity (becomes intradural) via foramen magnum (the seventh foramen traveled through) to join with the other vertebral artery to create the basilar artery.

Vertebral arteries have numerous branches and supply blood to a significant portion of the lower brainstem (medulla and pons) and the cerebellum. Vertebral arteries are susceptible to dissection secondary to neck trauma because of how they travel between the cervical vertebrae. They are essentially tethered while they traverse the transverse foramina leaving the intervertebral parts susceptible to dissection. Vertebral artery dissection can lead to either complete occlusion of the vertebral artery, producing ischemia, or artery to artery embolization to distal smaller branches, causing more focal symptoms.

This is shown in Figure A.3.

63

• • • • •

Off I branch at acute angle,
One of three to little brain,
Thrombus gives same side face numbness,
Horner's and hoarseness you'll gain.

Hint #1:

Wallenberg knows the answer.

Hint #2:

Not the eating of nonfood items, but spelled the same.

Posterior Inferior Cerebellar Artery

The posterior inferior cerebellar artery (PICA) branches from the V4 segment of the vertebral artery at an acute downward angle on its way to help perfuse the cerebellum (meaning "little brain" in Latin) along with the anterior inferior cerebellar artery and the superior cerebellar artery. PICA is the largest branch of the vertebral artery.

It courses anterior and then lateral to the medulla after its origin, giving off medullary branches. It then turns posteriorly, near the cerebellar tonsil, making an initial downward loop and then an upward loop before splitting into terminal cortical cerebellar branches.

This artery nourishes the large posterior and inferior surfaces of the cerebellum, inferior vermis, inferior cerebellar peduncle, and dorsolateral surface of the medulla.

Occlusion of the PICA may lead to an infarction in the entire vascular territory, including the medulla and the cerebellum or a limited infarction restricted to the lateral medulla known as the Wallenberg syndrome.

Typical findings in the Wallenberg syndrome include ipsilateral facial numbness to pain and temperature (because of lesion of the spinal trigeminal nucleus and tract), contralateral body numbness to pain and temperature (because of lesion of the spinothalamic tract), hoarse voice and dysphagia (because of lesion of nucleus ambiguus), and ipsilateral Horner's syndrome (miosis, ptosis, and anhidrosis) from damage to the descending sympathetic fibers.

Other important symptoms are ipsilateral ataxia and vertigo (because of the involvement of the inferior cerebellar peduncle and vestibular nuclei).

A complete infarction of the posterior inferior cerebellum can manifest as a life-threatening emergency because of edema in the posterior fossa, leading to transtentorial herniation and occlusion of the fourth ventricle, causing obstructive hydrocephalus. This usually requires emergent neurosurgical intervention such as a suboccipital craniectomy with or without an external ventriculostomy catheter placement.

This is shown in Figure A.3.

64

• • • • •

I run midline, flow front to back,
My floor a curv'ed blade,
Granulations supplement my flow,
Headache if clot is made.

Hint #1:

The most superior blood flow.

Hint #2:

Venous.

Superior Sagittal Sinus

The superior sagittal sinus is a key component of the dural venous sinus drainage system of the brain. It is positioned midline over the interhemispheric fissure and drains venous blood anterior to posterior, ending in the occipital area, where it meets the transverse sinuses at the confluence of sinuses (known as the torcular Herophili). At the torcular, the venous blood from the superior sagittal sinus drains into the right transverse sinus and the blood from the straight sinus drains into the left transverse sinus.

A commonly present large cortical vein known as the superior anastomotic vein of Trolard drains into the superior sagittal sinus. The inferior anastomotic vein of Labbe drains into the transverse sinus.

The floor of the superior sagittal sinus is the falx cerebri, which gets its name due to its shape, as *falx* in Latin means "curved blade" or "sickle." Within the superior sagittal sinus are projections called arachnoid granulations, which help drain the CSF into the venous system. The superior sagittal sinus is a common location for dural venous sinus thrombosis, which presents as an intense headache with signs of increased intracranial pressure such as vomiting and optic disc edema.

These patients can also develop venous infarcts due to venous congestion. Venous infarcts have a high propensity for secondary hemorrhagic transformation. The treatment of choice is anticoagulation.

This is shown in Figure A.4.

65

• • • • •

Vertically, and between the flow,
I stretch and separate,
The left and right and attach up front,
At the rooster and back at the straight.

Hint #1:

Some say I'm a sickle-shaped structure.

Hint #2:

Crista galli means "crest of the rooster" in Latin.

Falx Cerebri

The falx cerebri is a midline, dural reflection, positioned vertically in the middle. It contains the superior sagittal sinus in its superior edge and the inferior sagittal sinus in its inferior edge (*between the flow*).

It stretches down into the interhemispheric fissure and *separates the left and right hemispheres*. It attaches to the skull anteriorly at the crista galli (meaning "crest of the rooster" in Latin) and posteriorly over the *straight sinus* at the tentorium cerebelli.

Common pathologies associated with the falx cerebri include meningiomas arising from the falx, superior sagittal venous sinus thrombosis, and subdural hemorrhage.

Falcine meningiomas can be asymptomatic or symptomatic depending upon the size and location along with anterior-posterior axis. When these are closer to the central sulcus, patients can experience sensory or motor symptoms, predominantly in the lower extremities because of the representation of the leg and the foot on the mesial surface of the brain.

66

· · · · ·

Though I'm composed of lesser wing,
I'm wrapped in all three mothers,
I carry only one nerve, you see,
And one vessel to front, no others.

Hint #1:

19th word is a hint as well.

Hint #2:

Middle cranial fossa.

Optic Canal

The optic canal is located in the lesser wing of the sphenoid. Its walls are coated by the dura. Its chief contents are the optic nerve and ophthalmic artery, both of which travel anteriorly to the orbit.

At the apex of the orbital cavity, the optic canal is located very close to the superior orbital fissure, and the contents of both cavities converge at the apex before entering the intracranial space. Pathologies affecting the orbital apex can present with a constellation of signs and symptoms related to the nerves and vessels traversing the optic canal (optic nerve and ophthalmic artery) and the superior orbital fissure (cranial nerves III, IV, VI, sympathetic fibers, and the ophthalmic division of trigeminal nerve). Orbital apex syndrome can be differentiated from cavernous sinus syndrome by the involvement of the optic nerve and relative sparing of the maxillary division of the trigeminal nerve. In addition, there can be retro-orbital pain and bulging of the eye because of mass effect in orbital apex syndrome.

Optic nerve sheath meningiomas arise from the arachnoid cap cells of the optic nerve sheath as it passes through the optic canal. Direct compression from the meningioma can lead to vision loss. Imaging of choice to detect these meningiomas is MRI with and without contrast with dedicated cuts through the orbits. Surgical decompression is the treatment of choice to prevent complete vision loss.

67

• • • • •

Lying transverse, and in crescent-shape,
I separate zone from zone,
Attached up front at the Greek *like bed,*
And in back at occipital bone.

Hint #1:

The Greek term for "bed-like" is *clinoid.*

Hint #2:

Think: horizontal structure.

Tentorium Cerebelli

The tentorium cerebelli is the second largest dural reflection in the skull, after the falx cerebri. It lies transverse, separating the occipital lobes from the cerebellum and providing the key landmark for separating the brain into the supratentorial and infratentorial brain regions. Though many would disagree, it is often described as "crescent-shaped." It attaches anteriorly at the clinoid processes (*clinoid* in Greek is "bed-like") and attaches at the back of the occipital bone.

Transverse venous sinuses are present in the lateral edge of the tentorium cerebelli. The posterior edge of the falx cerebri terminates over the superior surface of the tentorium cerebelli, and this space contains the straight sinus.

The notch-shaped opening between the tentorial edges, which forms the only communication between supratentorial and infratentorial compartments is the tentorial incisura. A vertical transtentorial herniation of the brainstem either upward or downward via the incisura can occur in the setting of mass lesions in the infratentorial compartment or the supratentorial compartment, respectively.

A subtype of transtentorial herniation, known as uncal herniation, is characterized by lateral herniation of the temporal lobe and the uncus beyond the margin of the tentorium pushing against the brainstem. This condition can compress the brainstem, especially midbrain, against the contralateral sharp edge of the tentorium, forming a "notch" in the cerebral peduncle of the midbrain, known as the Kernohan–Woltmann notch phenomenon. Pressure on the corticospinal tract in the cerebral peduncle can lead to weakness ipsilateral to the original pathology that led to the herniation (false lateralizing sign) and oculomotor palsy.

This is shown in Figure A.1.

68

• • • • •

A continuation of the innermost layer,
Travel north to the south and in space,
Stretching from cone to coccygeal bone,
The *last thread* keeping taut and in place.

Hint #1:

The innermost dural layer that is.

Hint #2:

Starts around L2 or L3.

Filum Terminale

The filum terminale is a thin layer of connective tissue, continuous with the *pia mater* (the innermost meningeal layer) that *travels inferiorly* through the spinal canal (after the spinal cord ends via the *conus medullaris* at around L1-2), through the *subarachnoid space*, to attach to the *coccyx*. Its major function is to keep the *spinal cord pulled taut*. In Latin, *filum terminale* means "terminal thread."

In some cases, the filum terminale has a fatty collection (lipoma of the filum terminale), which is usually asymptomatic. In rare cases, it has been implicated in the pathogenesis of tethered cord syndrome. This is a pediatric neurological syndrome characterized by progressive lower extremity weakness and bowel and bladder incontinence. This is usually associated with a closed spinal dysraphism, including lipoma of the filum terminale or a tight filum. These patients require surgical intervention.

69

• • • • •

A combination of two layers,
Made up of irregular tissue,
I cover the brain and spinal cord,
And toughness is never my issue.

Hint #1:

Well, a bilayer in the skull, and only one layer over the spinal cord.

Hint #2:

One of my layers adheres to the skull.

Dura Mater

The dura mater is the outermost layer of the three meningeal layers and is composed of a bilayer of irregular connective tissue. The external layer is the periosteal layer, which is attached to the inner surface of the skull. The inner meningeal layer is typically fused with the external periosteal layer, except in places where the meningeal layer forms folds like the falx cerebri and the tentorium cerebelli.

Located in the skull and within the spinal canal, it surrounds the brain and spinal cord. *Dura mater* means "tough mother" in Latin.

The potential space between the periosteal layer of the dura and the inner surface of the skull is known as the epidural space. The middle meningeal artery (a branch of the external carotid artery) enters the skull via the foramen spinosum and travels in the epidural space. Epidural hematomas are caused by laceration of this artery.

The potential space between the inner meningeal layer and the arachnoid mater is known as the subdural space. This space is a very common site of bleeding, known as subdural hematoma. There are small bridging veins that traverse the subdural space to enter the dural venous sinuses. These veins are implicated in the pathogenesis of the subdural hematomas. The most common cause is head trauma, but in older patients or patients with significant brain atrophy, trivial head trauma can also produce a subdural hematoma because of stretching of the bridging veins as the brain parenchyma shrinks away from the dura.

Meningiomas are the most common intracranial benign tumors and the majority of them are intradural. A common imaging finding on T1 MRI with contrast is to identify a "dural tail" for these extra-axial tumors.

70

• • • • •

I'm the middle of three wrappers,
Named for my web-like tissue,
Atop me lies potential space,
Under me filled and true.

Hint #1:

Covers brain and spinal cord.

Hint #2:

Do not get swept away in the current that lies just under me.

Arachnoid Mater

Arachnoid mater is the middle meningeal layer between the dura and the pia mater. Its name is related to its web-like appearance and derived from the Greek word *arakhne* meaning "spider."

The potential space between the arachnoid layer and the dura mater is called the subdural space, as discussed earlier.

The space between the arachnoid mater and the pia mater is known as the subarachnoid space. This space contains the CSF, arteries, veins, and cranial nerves.

Near the dural venous sinuses, small projections of the arachnoid mater into the dural venous sinus through the dura mater can be seen. These are referred to as arachnoid granulations or Pacchionian granulations. These granulations allow the CSF to be drained into the venous sinuses. These granulations are mostly seen in the superior sagittal sinus and the transverse sinus.

71

• • • • •

The space between two Latin moms,
Where fluid runs, but if,
Headache and fever, be concerned,
And tap if neck gets stiff.

Hint #1:

Just below a like-named structure.

Hint #2:

Where you might find white blood cells, proteins, bacteria, glucose, and red blood cells.

Subarachnoid Space

The subarachnoid space is a true space *between the pia mater and the arachnoid mater* (*mater* in Latin means "mother"). It is filled with CSF and is the site of infection in cases of meningitis, which often presents with *headache, fever, and neck stiffness.* Meningitis is diagnosed by lumbar puncture, also called a *spinal tap.*

In addition to the CSF, the subarachnoid space also contains all the major arteries and cranial nerves as they exit or enter the brain. Multiple cranial nerve palsies can be a feature of infectious or inflammatory conditions involving the CSF in the subarachnoid space, for example, infectious meningitis or carcinomatous meningitis.

Subarachnoid hemorrhage due to a ruptured aneurysm of a cerebral artery is a life-threatening neurological emergency that presents as sudden onset, severe headache with nausea, vomiting, and loss of consciousness. The two major complications in these patients are delayed cerebral ischemia (due to vasospasm) and hydrocephalus (because of decreased reabsorption of CSF via arachnoid granulations or due to extension of the hemorrhage into the ventricles).

72

• • • • •

The closest to the gray matter,
I adhere, and invaginate,
To allow blood vessels through choroid plexus,
I help create.

Pia Mater

The pia mater is the innermost layer of meninges that adheres directly to the surface of the brain, invaginating into its folds and sulci, and is highly vascularized with leptomeningeal blood vessels. Together with the arachnoid mater, these layers are referred to as leptomeninges.

The pia mater is separated from the arachnoid mater by the CSF-containing subarachnoid space. As the blood vessels traversing the subarachnoid space enter the parenchyma of the brain, they pierce the pia mater and often carry with them a sleeve of pia mater with CSF for a short distance. These perivascular spaces are known as the Virchow–Robin (VR) spaces.

The choroid plexus, present in the ventricles, is responsible for CSF production. The plexus consists of highly vascularized connective tissue, known as tela choroidea, which is a derivative of pia mater.

73

• • • • •

Separating two lobes from one,
Positioned diagonally,
A cleft on left as well as right,
Open, insula you'll see.

Hint #1:

Just medial to the most lateral parts of the brain.

Hint #2:

Angled postero-superiorly to antero-inferiorly.

Lateral Sulcus/Sylvian Fissure

The lateral sulcus/sylvian fissure separates the temporal lobe from the frontal and parietal lobes. It is a bilateral structure, which angles diagonally, postero-superiorly to antero-inferiorly, and if opened slightly makes the insular lobe visible.

It consists of three rami – anterior horizontal, anterior ascending, and posterior. The anterior horizontal ramus separates the pars orbitalis from the pars triangularis. The anterior ascending ramus separates the pars triangularis from the pars opercularis.

The posterior ramus is the longest and extends to the end of the sylvian fissure that is surrounded by the supramarginal gyrus.

The CSF-filled space within the sylvian fissure is referred to as the sylvian cistern. The anterior part of this cistern separates the orbitofrontal lobe from the temporal lobe and communicates with the carotid cistern medially.

Posteriorly, the cistern can be divided into an insular cleft and an opercular cleft. The insular cleft separates the insula from the frontoparietal and temporal operculum and forms the circular sulcus of the insula. It contains the M2 segment of the middle cerebral artery.

The opercular cleft is located lateral to the insular cleft and contains the M3 segment of the middle cerebral artery.

An important anatomical landmark used in epilepsy surgery is the junction of sylvian fissure and the central sulcus. In the language-dominant hemisphere, the Wernicke's area is situated ~1 cm behind this junction and is mostly restricted to the superior temporal gyrus.

74

• • • • •

I lie below a like-named pair,
And send chemicals down a stalk,
Weight gain, fatigue, and low libido,
If from me the chemicals stop.

Hint #1:

Think of the adrenal axis.

Hint #2:

Secretes thyroid-releasing hormone and gonadotropin-releasing hormone (among several others).

Hypothalamus

The hypothalamus is a derivative of diencephalon and is situated below the thalami (*like-named pair*). The hypothalamus is a collection of nuclei with various important functions.

The most medial nuclei are the periventricular nuclei, situated immediately next to the third ventricle. Then there is a medial hypothalamic group and a lateral hypothalamic group. In addition, these nuclei are organized into four groups along the anterior-posterior axis: the preoptic area, supraoptic region, tuberal region, and mammillary region.

The lateral hypothalamic group contains the lateral hypothalamic nucleus and the medial forebrain bundle (MFB) with some other smaller nuclei. The medial hypothalamic group includes the medial preoptic nucleus, paraventricular nucleus, supraoptic nucleus, suprachiasmatic nucleus, arcuate nucleus, dorsomedial nucleus, ventromedial nucleus, mammillary nuclei (medial and lateral), and posterior hypothalamic nucleus.

The suprachiasmatic nucleus plays an important role in the regulation of circadian rhythm. The ventrolateral preoptic nucleus (VLPO) is important for promoting sleep. On the other hand, the posterior hypothalamic nucleus produces orexin, which promotes wakefulness. Hence a lesion of the anterior hypothalamus can cause insomnia, whereas a posterior hypothalamic lesion can lead to hypersomnia. Also, anterior hypothalamus promotes heat dissipation, whereas posterior hypothalamus promotes heat conservation.

The mammillary nuclei are part of the Papez circuit and receive information from the hippocampus via the fornix and then project to the thalamus via the mammillothalamic tract.

In addition, arcuate, periventricular, medial preoptic, and paraventricular nuclei produce release and inhibitory factors that modulate the production of pituitary hormones. Oxytocin and vasopressin, released by posterior pituitary, are produced by supraoptic and paraventricular nuclei, respectively.

A very peculiar presentation of a benign tumor involving the hypothalamus (hypothalamic hamartoma) is a gelastic seizure. These are seizures that are characterized by stereotypical attacks of laughter. These children can also have precocious puberty. It is very important to recognize the gelastic seizures and assess for a hypothalamic hamartoma using a thin-cut MRI of brain with sella view. Early treatment of the hamartoma can prevent progression into a developmental epileptic encephalopathy.

75

● ● ● ● ●

One cave each side of the Turkish saddle,
Five nerves, one vessel within,
If thrombus creates a nervous palsy,
Your eye first starts to look in.

Hint #1:

The Latin *sella turcica* means "Turkish saddle."

Hint #2:

Also think eye swelling and pain.

Cavernous Sinus

The cavernous sinus is a dural venous sinus present in the base of the skull. There is a pair of cavernous sinuses, one on each side of the sella turcica (Latin meaning "Turkish saddle"), which contains the pituitary gland.

The cavernous sinus contains fibrous septae, which give it the appearance of a "cavern." Longitudinally, the cavernous sinus extends from the orbital apex to the petrous apex.

Anteriorly, the sinus receives venous blood from superior and inferior ophthalmic veins, sphenoparietal sinus, intercavernous sinus, and the superficial middle cerebral vein.

Posteriorly, the cavernous sinus drains into the transverse sinus via the superior petrosal sinus and the jugular bulb via the inferior petrosal sinus.

Some important structures present within the cavernous sinus include the internal carotid artery, surrounded by the sympathetic plexus. Immediately adjacent to the artery is the abducens nerve. In the lateral wall of the cavernous sinus, cranial nerves II, III, IV, V1, and V2 are present (superior to inferior).

Pathologies invading the cavernous sinus, for example, tumor, cavernous sinus thrombosis, carotico-cavernous fistula, and so on often present with symptoms of headache, proptosis, and conges-tion of the eye and any combination of cranial neuropathies involving the above-mentioned nerves. This constellation of signs and symptoms is referred to as cavernous sinus syndrome. An incomplete cavernous sinus syndrome might just include Horner's syndrome and abducens nerve palsy causing abduction paresis (*eye first starts to look in*). This is known as Parkinson's syndrome.

Any form of cavernous sinus syndrome is a neurological emer-gency and warrants emergent imaging with MRI of brain, MRI orbits with and without gadolinium, along with venogram of brain.

76

• • • • •

I drain from lateral,
To most superior of veins,
I poke through the tough dura,
Crescent-shaped bleed atop your brain.

Hint #1:

Most commonly ruptured with head trauma (elderly fall).

Hint #2:

Vein? More specifically a sinus.

Bridging Veins

The bridging veins drain venous blood from the *brain medially into the superior sagittal sinus,* which is the *most superior venous pathway* in the brain. They *poke through the dura mater* to do so and are often sheared in the case of head trauma.

Cerebral veins and arteries travel along with cranial nerves in the subarachnoid space, a CSF-filled space between the arachnoid mater and the pia mater. Some veins pierce the arachnoid mater and the overlying dura mater to drain directly into the dural venous sinuses. These are bridging veins. As they pierce the arachnoid and the dura, they travel through the "potential" subdural space.

Severe head trauma can lead to the rupture of these veins with consequent bleeding into the subdural space, known as the subdural hematoma. This is a very common form of intracranial bleeding, requiring emergent admission to hospital and often neurosurgical intervention.

A subdural hematoma appears *crescent-shaped on a non-contrasted CT scan,* crossing the suture lines.

77

• • • • •

Five roots that intertwine and blend,
On way from the cervix,
Injured with fall or birthing that,
Klumpke and Erb cannot fix.

Hint #1:

The spaghetti of the axilla.

Hint #2:

Think dermatomes and myotomes.

175

Brachial Plexus

The brachial plexus is composed of *five nerve roots* (C5–T1), which exit the cervical spine (except for T1), *cross and combine* in many locations, and travel laterally to the arm.

The exiting nerve roots join to form trunks. C5 and C6 join to form the upper trunk, C7 continues as the middle trunk, and C8 and T1 join to form the lower trunk. Each trunk has an anterior and a posterior division.

The anterior divisions of the upper and middle trunks join to form the lateral cord. The anterior division of the lower trunk continues as the medial cord. The posterior divisions of all trunks join to form the posterior cord.

The lateral cord and medial cord join to form the median nerve, and the lateral cord then continues as the musculocutaneous nerve. The medial cord continues as the ulnar nerve, and the posterior cord forms the radial nerve.

This plexus is susceptible to injury, especially at two locations. The C8–T1 roots are often *injured during falls* when the patient is attempting to grab onto something to break their fall, or *during childbirth* if the arm exits first and is pulled, producing *Klumpke palsy*. Injury to C5–C6, an *Erb's palsy*, can also occur during childbirth when the neck is side-bent too far, or the shoulder is depressed.

Parsonage–Turner syndrome, aka neuralgic amyotrophy, is an idiopathic inflammatory condition involving the brachial plexus. It typically presents with severe pain lasting for a few weeks, followed by weakness and atrophy of muscles in the distribution of the upper and middle brachial plexus. It often affects the muscles innervated by the suprascapular, musculocutaneous, long thoracic, and radial nerves. Scapular winging is often seen on exam.

78

• • • • •

Named for my Latin shaggy hair,
Through the outer wall I grow,
Allowing only one-way travel,
From space to venous flow.

Hint #1:

Latin word *villus* means "shaggy hair."

Hint #2:

Blockages can create hydrocephalus.

Arachnoid Villi/Arachnoid Granulations

Arachnoid Villi/Arachnoid Granulations, aka Pacchionian granulations, are herniations of the arachnoid membrane that *project through the dura mater*, getting their name villi from the Latin word *villus* meaning "shaggy hair."

The function of the arachnoid granulations is to drain CSF *from the subarachnoid space into the dural venous sinuses*, while *preventing retrograde flow* of venous blood into the subarachnoid space.

Pathology affecting the leptomeninges (pia and arachnoid mater) can affect the drainage ability of arachnoid granulations, leading to the development of hydrocephalus. This is a common occurrence in bacterial meningitis and subarachnoid hemorrhage, during which pus or blood products can block the arachnoid granulations and lead to hydrocephalus.

79

• • • • •

Living in the pons I work,
To keep you safe and alert,
First named for my blue melanin,
I help produce adrenaline.

Hint #1:

Also called the nucleus pigmentosus pontis.

Hint #2:

The Latin word *ceruleus* means "of or pertaining to the sea or sky."

Locus Coeruleus

The locus coeruleus is a nucleus located in the dorsorostral pons and functions primarily to produce norepinephrine/noradrenaline. It is also called nucleus pigmentosus pontis, discovered by Felix Vicq-d'Azyr in the 1700s and gets its name from Latin, meaning "blue spot," with the Latin *ceruleus* meaning "of or pertaining to the sea or sky," that is, blue, and *locus* meaning "place," as unstained brain tissue appears to have a blue or azure hue.

The locus coeruleus is the chief source of norepinephrine in the brain and thus contains many noradrenergic neurons. Some noradrenergic neurons are also present in the lateral tegmental region in caudal pons.

The transition from sleep to wakefulness relies on a synchronized dance of various neurotransmitters, the chief of which is orexin, released by the posterior hypothalamus. Wakefulness, additionally, requires acetylcholine from the pedunculopontine nucleus and norepinephrine from the locus coeruleus.

The locus coeruleus is also the site of early or late involvement in various neurodegenerative pathologies like alpha-synucleinopathies. Early involvement in multiple system atrophy leads to early dysautonomia as opposed to the relatively late onset of autonomic dysfunction in idiopathic Parkinson's disease.

80

• • • • •

We are a group, and not just one,
Named for the midline *seam* we run,
Help regulate sleep–wake cycle,
We help produce 5-HT1.

Hint #1:

Think brainstem.

Hint #2:

Key component in the formation of our natural circadian rhythm.

Raphe Nuclei

The raphe nuclei are a group of nuclei clustered in the brainstem that get their name from the Greek word for "seam" due to their midline position. The more rostral group of raphe nuclei have extensive cortical projections and play a role in sleep–wake regulation, modulation of behavior, and emotion. The more caudal nuclei project mainly to the spinal cord and play a significant role in pain modulation.

The raphe interpositus is a special midline nucleus in the pons, which plays an important role in saccade generation. It contains the so-called omnipause neurons, which are tonically active and inhibit the activity of excitatory and inhibitory burst neurons to prevent unnecessary saccade generation. In order to generate a voluntary saccade, the descending input from the superior colliculus releases the omnipause neurons, which then activates the excitatory burse neurons.

The raphe interpositus has been implicated in diseases with abnormal saccades such as spinocerebellar ataxia 3 and progressive supranuclear palsy.

81

· · · · ·

A cleft between wings both great and small,
Through which travels nerves to the ball,
Three, four, six and branches of V1,
As well as two vein and arteries none.

Hint #1:

Ball aka globe.

Hint #2:

Not a "foramen."

Superior Orbital Fissure

The superior orbital fissure is a cleft, not a foramen, positioned at the back of the orbital cavity with its two largest opposing borders being the lesser wing of the sphenoid bone superiorly and the greater wing of sphenoid bone inferiorly. Through this fissure travel cranial nerves III, IV, VI, and the ophthalmic branch of the trigeminal nerve (more specifically the lacrimal and frontal nerves that branch from the V1), as well as the superior and inferior branches of the ophthalmic vein.

Tumors or other infiltrating pathologies invading the superior orbital fissure present with eye pain, proptosis, congestion, and a variable combination of cranial neuropathies affecting the above-mentioned cranial nerves. This clinical presentation is called superior orbital fissure syndrome because of its unique localization. Often, these patients also have associated optic neuropathy because of the involvement of the optic canal, which runs very close to the superior orbital fissure. This presentation is called orbital apex syndrome.

82

• • • • •

Nerve roots will pass right down my side,
Through true and fluid-filled space,
From tender mother to the dura,
Our job, the cord to brace.

Hint #1:

Triangular structures.

Hint#2:

Positioned horizontal to the spinal cord.

Denticulate Ligaments

The denticulate ligaments are triangular extensions of the pia mater (Latin meaning "tender mother"), which project laterally through the subarachnoid space from the surface of the spinal cord to the dura mater. The primary function of the denticulate ligaments is to brace the spinal cord in place within the dural sac.

Johann Jacob Huber described these ligaments first in 1739. Multiple authors have described the shapes and anatomy of these ligaments differently. But in a detailed anatomical study, it was observed that these ligaments are only present in the cervical and thoracic regions.

These ligaments separate the spinal canal into an anterior compartment and a posterior compartment. The ligaments in the cervical region have more collagen fibers compared to the ones in the thoracic region.

For surgical interventions on the spine, these ligaments serve as significant landmarks to identify the midline of the spinal cord.

83

. . . . .

Formed by the borders of two separate bones,
I'm positioned at the base,
Through me travel three nerves, and one vein,
Swallow, shrug shoulders, slow pace.

Jugular Foramen

The jugular foramen is a large cavity positioned at the base of the skull, with its posterior border being the occipital bone and its anterior border being the petrous portion of the temporal bone. Through its aperture passes the glossopharyngeal (innervates the swallowing muscles), vagus (provides parasympathetic innervation to the heart), and accessory nerves (innervates muscles that allow for shrugging shoulders) as well as the jugular vein.

Skull-based tumors at or near the jugular foramen can present with cranial neuropathies involving the glossopharyngeal, vagus, and spinal accessory nerves. This clinical presentation is called jugular foramen syndrome.

The jugular foramen is often the site of a type of a paraganglioma known as glomus jugulare, typically located along the glossopharyngeal and vagus nerves. The currently used term is jugulotympanic paraganglioma. As opposed to more common paragangliomas like pheochromocytomas, glomus jugulare tumors do not secrete catecholamines, and their clinical presentation is mostly related to compression of the cranial nerves at the jugular foramen. The tumor can also extend into the middle ear causing hearing loss and tinnitus. On otoscopic examination, a red pulsatile mass can be seen behind the tympanic membrane.

84

• • • • •

Descend, both sided, adjac'd the heart,
Come from three roots above,
Sensation to the organs' coating,
And provides the breath we love.

Hint #1:

Innervates to the pericardium, as well as mediastinal and diaphragmatic pleura.

Hint #2:

Motor innervation of the diaphragm.

Phrenic Nerve

The phrenic nerves originate from nerve roots C3–C5 and, after formation from these roots, descend bilaterally through the thorax, anterior to the root of the lung, and pass adjacent to the heart on the way to the diaphragm. The phrenic nerves provide sensory innervation to the pericardium, mediastinal and diaphragmatic pleura, and central tendon of the diaphragm, as well as motor innervation to the diaphragm.

The key neuroanatomical structures controlling respiration are located in the medulla. Most of these nuclei are located in the ventrolateral medulla (ventral respiratory group), but there is a dorsal respiratory group as well.

The ventral group contains the Botzinger complex, pre-Botzinger complex, rostral ventral respiratory group, and caudal ventral respiratory group in a rostro-caudal manner. The pre-Botzinger complex is considered to be the chief central pattern generator for respiration.

The inspiratory neurons primarily project to the C3–C5 spinal cord to control the phrenic nerve, and the expiratory neurons project to thoracic spinal cord segments.

85

· · · · ·

A hole in the back-most fossa,
That two nerves, one art, pass through,
If tumor were to compress it,
Cannot walk straight, feel face, hear you.

Hint #1:

A laterally oriented passageway.

Hint #2:

Entry and exit are common location for acoustic neuromas/vestibular schwannomas.

Internal Acoustic Meatus

The internal acoustic meatus is a foramen located in the posterior cranial fossa through which the facial nerve, vestibulocochlear nerve, and labyrinthine artery pass. Near the entry and exit of this canal is a common location for tumors called acoustic neuromas/vestibular schwannomas, which produce dysfunction of the facial nerve and vestibulocochlear nerves, for example, imbalance and hearing loss.

The internal acoustic canal is divided into a superior and an inferior compartment by the falciform crest. The superior compartment contains the facial nerve and the superior vestibular nerve. The inferior compartment contains the inferior vestibular nerve and the cochlear nerve. The vestibular ganglion (of Scarpa) is also located within the canal.

This rigid bony canal is implicated in the pathogenesis of Bell's palsy. As the nerve becomes inflamed, the edematous nerve gets compressed in the bony canal, further compromising blood flow.

86

• • • • •

The largest in the skull I sit,
In the lowest of its part,
Through me passes eleven,
The brainstem and five arts.

Hint #1:

The most caudal portion of the brainstem passes through this.

Hint #2:

Two of the arteries that pass through have the same name.

Foramen Magnum

The foramen magnum is the bottom-most foramen in the skull and the largest. Through it passes the spinal accessory nerve (CN XI) as well as five arteries: the two vertebral arteries, the posterior spinal arteries, and the anterior spinal artery. The inferior medulla also passes through this opening to join the cervical spinal cord.

The CSF-filled cistern at the cervicomedullary junction near the foramen magnum is called the cisterna magna. At this level, most of the corticospinal tract fibers decussate (pyramidal decussation). The decussating fibers have a somatotopic organization, with the arm fibers being medial and the leg fibers being lateral. In addition, the arm fibers decussate slightly higher than the leg fibers. This peculiar arrangement can lead to unusual patterns of weakness with pathologies affecting the cervicomedullary junction near the foramen magnum.

Cruciate paralysis refers to bilateral arm weakness with relative sparing of legs. This would indicate the presence of a midline lesion. Hemiplegia cruciate refers to arm weakness with contralateral leg weakness, which can occur because of a lateral lesion affecting the decussated arm fibers and the leg fibers pre-decussation.

87

• • • • •

Posed behind the ICA,
Near axis and supplying,
Autonomics to the head,
Big pupils and stop crying.

Hint #1:

"Axis" comes from the Latin word for *axil,* and another name for C2.

Hint #2:

Also helps with salivation and goosebumps.

Superior Cervical Ganglion

The superior cervical ganglion is formed by the fusion of paravertebral ganglia in the cervical region C1–C4. It is located at the level of C2 and C3 vertebrae and just behind the internal carotid artery, just after it branches from the common carotid artery.

The first-order sympathetic neurons descend from the hypothalamus laterally within the brainstem and synapse in the intermediolateral column with the ciliospinal center of Budge at the C8–T2 levels.

The second-order neuron then ascends above the lung apex via the sympathetic trunk and reaches the superior cervical ganglion, where it synapses with the third-order neuron.

Finally, these axons ascend along with the common carotid artery. The fibers destined for the sweat glands in the face separate and travel with the external carotid artery. The rest of the fibers ascend along with the internal carotid artery to reach inside the skull. These fibers are destined for the dilator pupillae muscle in the eye, the superior and inferior tarsal muscles, and the lacrimal gland.

A lesion anywhere along this pathway can produce Horner's syndrome, characterized by miosis and ptosis with or without anhidrosis.

88

• • • • •

I live just under the diaphragm,
From Greek *abdomen*, I'm named,
Secrete, feel pain, start churning I signal,
Injected if always inflamed.

Hint #1:

Innervates the foregut.

Hint #2:

Common location for a "nerve/plexus block."

Celiac Plexus/Ganglion

The celiac plexus/ganglion is located just inferior to the diaphragm in the retroperitoneal space. It lies anterolateral to the aorta at the level of the celiac trunk.

Celiac in Greek means "abdomen." This plexus functions to provide visceral sensory transmission to/from the organs of the foregut (namely gastric secretion), sense visceral pain, and promote peristalsis/digestion. This is also a common location for nerve blockade (celiac plexus block) in the case of recurrent and refractory upper abdominal pain.

The plexus is created by preganglionic sympathetic fibers in the splanchnic nerves and the preganglionic parasympathetic fibers from the vagus nerve.

89

• • • • •

Around a like-named artery,
I sit retro and midline,
Give movement to the midgut,
And help digest once you dine.

Hint #1:

A network or hub of nerves.

Hint #2:

A very sympathetic region.

Superior Mesenteric Plexus/Ganglion

The superior mesenteric plexus/ganglion is positioned retroperitoneal atop the abdominal aorta at the level of the superior mesenteric artery. Its major function is innervating the organs of the midgut, namely the jejunum, ileum, and ascending colon.

The branches of the celiac plexus and posterior vagal trunk form the superior mesenteric plexus.

The preganglionic sympathetic neurons are located in the intermediolateral column of the thoracolumbar spinal cord. The two types of sympathetic ganglia are paravertebral ganglia, like the superior cervical ganglion and the prevertebral ganglia, like the celiac ganglion and the superior mesenteric ganglion.

90

• • • • •

Over a like-named vessel,
I sit upon the Greek *to hang,*
Though some may call me "lesser than,"
I feel two-thirds your colon pang.

Hint #1:

Aorta comes from the Greek word *aortéō* meaning "to hang" or "to lift."

Hint #2:

Innervates the smooth muscle of the hindgut.

Inferior Mesenteric Plexus

The inferior mesenteric plexus sits just superior to the inferior mesenteric artery, positioned on top of the aorta, which in Greek means "to hang" or "to lift." Its primary function is providing sensory and autonomic innervation to the distal two-thirds of the large intestine/colon.

As opposed to the superior mesenteric plexus, the inferior mesenteric plexus is not formed by branches of the celiac plexus. Instead, the preganglionic sympathetic neurons for this plexus emerge from L1–L3 spinal levels via the splanchnic nerves.

For refractory lower abdominal pain, CT-guided inferior mesenteric ganglion block can be a useful treatment modality.

91

. . . . .

I start at a close like-named horn,
And exit lat alone,
No ganglion, through foramen,
To move muscle of bone.

Hint #1:

The front-most horn that is.

Hint #2:

No sensation here.

Ventral/Anterior Nerve Root

The anterior/ventral nerve root originates at the anterior horn. After exiting the spinal canal laterally, it passes through the intervertebral foramen and, unlike the posterior/dorsal nerve root, it does not pass through a ganglion.

The ventral root contains the axons of the alpha motor neurons from the anterior horn and the axons of the preganglionic sympathetic neurons originating from the intermediolateral cell column.

The motor axons in the ventral root join the sensory axons in the dorsal root to form the mixed spinal nerve. The preganglionic sympathetic axons also join the mixed nerve for a short distance and then exit via the white ramus to enter the paravertebral ganglion, and the postganglionic axons re-enter the nerve via the gray ramus.

92

• • • • •

Coming in afferently,
Dorsal horn from bony hole,
In ganglion, nerves do not synapse,
Your feelings are my role.

Hint #1:

No motor here.

Hint #2:

The root of your pain.

Dorsal/Posterior Nerve Root

The dorsal/posterior nerve root is an afferent nerve that brings in sensory information from the periphery to the spinal cord. The dorsal root ganglion, containing the cell bodies of the sensory afferent fibers, is located outside the spinal canal. These fibers then travel in the posterior root and enter the spinal cord via the dorsal horn, where they synapse with neurons in various laminae (substantia gelatinosa) before ascending either in the dorsal columns ipsilaterally (touch, vibration, and proprioception) or in the spinothalamic tract contralaterally (pain and temperature).

In patients with radiculopathy, nerve conduction studies often show completely normal SNAPs, even though patients complain of sensory symptoms. This is because the dorsal root ganglia are located outside the spinal cord, and pathology in radiculopathy is proximal to these ganglia. When nerve conduction studies are performed on peripheral nerves, fibers distal to the ganglia (which are intact) are stimulated, and hence SNAPs remain normal.

Sensory neuronopathies or ganglionopathies are diseases that affect the dorsal root ganglia selectively and produce a pure sensory syndrome characterized by numbness, tingling, decreased reflexes, and ataxia. On nerve conduction studies, the SNAPs are quite abnormal with decreased amplitudes with relatively preserved compound muscle action potential (CMAP). Some important causes include Sjogren's syndrome, vitamin B6 toxicity, and paraneoplastic processes with anti-Hu antibodies.

93

• • • • •

I sit in back under the splenium,
Many call me *third eye*,
I help to regulate your sleep,
With age I calcify.

Hint #1:

Interestingly endocrine.

Hint #2:

Highly sensitive to light.

Pineal Gland

The pineal gland is a small endocrine gland positioned posterior to the third ventricle and inferior to the splenium of the corpus callosum. Throughout history, many mystics and psychics have called it the "third eye," while others give it this name because of how sensitive it is to light. The major function of the pineal gland is to produce melatonin in response to light or darkness in the environment. Calcification of the pineal gland is common in older age, and often called "brain sand."

The pineal gland sits immediately above the tectum (quadrigeminal plate) of the midbrain, especially the superior colliculi. Pineal gland tumors often present clinically with dorsal midbrain syndrome (Parinaud syndrome). This is characterized by supranuclear gaze palsy, bilateral lid retraction (Collier's sign), light-near dissociation, and convergence–retraction nystagmus.

This is shown in Figure A.1.

94

• • • • •

A nucleus in basal forebrain,
Some call me Meynert,
My cholinergic neurons help
Remember names and stay alert.

Hint #1:

I project to the cerebral cortex and the hippocampus.

Hint #2:

It receives input from the substantia nigra (dopaminergic), the raphe nuclei (serotonergic), and the nucleus locus coeruleus (norepinephrine/noradrenaline).

Nucleus Basalis

The nucleus basalis is a nucleus located in the basal forebrain, also known as the nucleus basalis of Meynert (NBM). This is a poorly defined area in the base of the forebrain and is thus also known as the substantia innominata.

This region is located above the optic nerve with the lateral ventricle forming the medial wall. Although this region is poorly defined, it is clear that there are different cholinergic nuclei present with different targets. The nucleus basalis of Meynert is one of these nuclei that projects to extensive regions of the cerebral cortex.

The other nuclei include the medial septal nucleus and the vertical limb of the nucleus of the diagonal band of Broca, both of which project to the hippocampus. The horizontal limb of the nucleus of the diagonal band of Broca projects to the olfactory bulb.

Atrophy of the basal forebrain cholinergic system, including that of the NBM, and subsequent reduction of cholinergic innervation of the cortex is thought to be an important part of the pathophysiologic process of Alzheimer's disease. Currently, studies analyzing the effects of deep brain stimulation (DBS) of the NBM on cognitive performance in patients with Alzheimer's disease are undergoing.

95

• • • • •

Most caudal in the brain stem,
Where motor nerves begin to slide,
From left to right and right to left,
One half controls the other side.

Hint #1:

Located in the medulla.

Hint #2:

Follow corticospinal tract.

Pyramidal Decussation

The pyramidal decussation is located in the inferior medulla, the lowest portion of the brainstem. Here, the axons of the corticospinal tract cross (decussate); this is the reason why the right brain controls the left half of the body, and the left brain controls the right. The anatomy of the corticospinal fibers in this region is complicated. The fibers destined for upper extremity are traveling more medially and anteriorly compared to those destined for lower extremity. In addition, the decussation of the upper extremity corticospinal fibers occurs above the level of the foramen magnum, and that of lower extremity fibers occurs inferiorly. This complex anatomy can lead to unusual clinical syndromes when pathologies affect this region.

Cruciate paralysis is characterized by bilateral upper extremity paralysis with relative sparing of lower extremities, and this localizes to the cervicomedullary junction. Lesions restricted to one side in this region can present with ipsilateral paralysis of the upper extremity and contralateral paralysis of the lower extremity. This clinical presentation is known as hemiplegia cruciata.

Congenital mirror movement (CMM) disorder is a rare disorder characterized by mirror movements of distal upper extremities. This is related to mutation of the *DCC* and *RAD51* genes, which play an important role in pyramidal decussation. Patients with CMM have an abnormal undecussated corticospinal tract; hence, signals generated in one primary motor cortex tend to activate bilateral muscle groups.

96

• • • • •

A branch of the PICA or vert,
Perfusing one-third in the back,
If I clot, vibration and
Position sense you'll lack.

Hint #1:

A paired structure.

Hint #2:

Run behind the columns.

Posterior Spinal Arteries

The posterior spinal arteries branch off the posterior inferior cerebellar artery in approximately 75% of people, and from the vertebral artery in the remaining 25%. They perfuse the posterior one-third of the spinal cord, specifically the dorsal columns, which provide vibration and special sense information from the periphery to the brain.

The posterior spinal arteries exist as a pair as opposed to a single anterior spinal artery. These arteries travel on the posterolateral surface of the spinal cord. Multiple other arteries, entering through the intervertebral foraminae, supplement this artery, one of them being the artery of Adamkiewicz between T9 and T12 levels.

Ischemia in the territory of this artery is very rare, given the posterior spinal arteries run in pairs with extensive connections between the two.

97

• • • • •

We sit, both sides, just in front
Of the precentral sulcus,
Why eyes are fixed to the same side,
When you infarct one of us.

Function in goal-directed eye movements and localizing visual targets.

Name = three words.

Frontal Eye Field

The frontal eye field (FEF) or Brodmann area 8 is situated in the posterior part of the middle frontal gyrus at the junction of the precentral sulcus and superior frontal sulcus. This cortical area is thought to play a critical role in the generation of eye movements (saccadic and smooth) toward the contralateral direction.

Primate studies have shown that this area contains visual neurons, movement neurons, and visuomovement neurons. Direct electrical stimulation studies of this area in both humans and nonhuman primates have elicited saccadic and smooth eye movements in the contralateral direction. This area has been found to have extensive connections to the superior colliculus, caudate nucleus, and to the parietal eye field.

Primate stimulation studies have additionally shown that the amplitude of saccades has a topographic organization within the FEF, with small-amplitude saccades represented in the ventro-lateral region and large amplitude saccades in the dorso-medial region.

When a stroke affects the frontal lobe including the FEF, patients often have a gaze deviation toward the lesioned hemisphere because of an imbalance between the activity of the two FEFs and the relative overactivity of the healthy hemisphere deviating the gaze toward the lesioned hemisphere.

On the other hand, in patients with seizures involving the FEF, forced sustained gaze and head deviation (defined as version) occurs in the direction contralateral to the hemisphere of seizure onset. It is a highly reliable lateralizing sign.

98

• • • • •

We sit within the midbrain,
And once light comes in on two,
Synapsing once, and then with us,
Constrict pupils we do.

Hint #1:

Post-pretectal nuclei.

Hint #2:

Autonomic function.

Edinger–Westphal Nucleus

The Edinger–Westphal nucleus, also called the accessory oculo-motor nucleus, is the primary provider of parasympathetic innervation to the sphincter muscles of the iris and to the ciliary muscles. It is located in the midbrain bilaterally and is a key component of the pupillary light reflex.

After light enters the eye through the pupil, the signal is transmitted by the optic nerve via the optic chiasm and then the optic tract to synapse at the bilateral pretectal nuclei. From there, fibers travel to and synapse at the Edinger–Westphal nucleus before traveling back toward the orbit via the oculomotor nerve.

These preganglionic parasympathetic nerve fibers travel with the oculomotor nerve and synapse at the ciliary ganglion from whence the postganglionic parasympathetic fibers arise. These fibers travel via the short ciliary nerves to constrict the pupillary sphincter and allow less light into the eye through the pupil.

From the level of the pretectal nuclei, this pathway is bilateral, so light stimulus presented to one eye not only activates the pupillary constriction in that eye but also in the contralateral eye (consensual light reflex).

In addition to light reflex, pupillary constriction can also occur as part of the near response. In pathologies affecting the dorsal midbrain (Parinaud syndrome), patients often have light-near dissociation, where the light reflex is abolished but the near response stays intact (Argyll Robertson pupil).

99

• • • • •

Descendants in the Latin bridge,
And of the seventh son,
On way to pass the hammer by,
Engaged 'til dinner's done.

Hint #1:

Latin word for "bridge" is *pontem/pons.*

Hint #2:

Latin name for "hammer" is *malleus,* which is also the name of a middle ear bone.

Chorda Tympani

The chorda tympani is a branch of the facial nerve, CN7, *the seventh son*, which itself originates at the pons, which in Latin means "bridge." Specifically, the chorda tympani branches off as the facial nerve passes through the mastoid bone posterior to the middle ear (mastoid segment of the facial nerve).

The chorda tympani has a complicated path as it travels anteriorly through the middle ear, passing between the incus and the malleus, just inside the tympanic membrane. Thus, it *passes the hammer (malleus in Latin) by*. From the middle ear, it passes antero-inferiorly through the petrotympanic fissure and into the infratemporal fossa where it joins the lingual nerve, a branch of the mandibular division of the trigeminal nerve (V3).

The lingual nerve carries the chorda tympani fibers to their destination. The special sensory afferent fibers carry taste sensation from the anterior two-thirds of the tongue, and thus it is *engaged 'til dinner's done.*

The preganglionic parasympathetic fibers from the superior salivatory nucleus are destined for the submandibular ganglion, which then gives off postganglionic parasympathetic fibers to the submandibular and sublingual salivary glands.

In most cases of peripheral facial nerve palsy, there is a loss of taste sensation in the anterior two-thirds of the tongue, except when the lesion is distal to the origin of the chorda tympani nerve in the middle ear. If the taste sensation is spared, salivation will be spared as well, since the parasympathetic fibers destined for the submandibular ganglion exit the middle ear along with the taste fibers in the chorda tympani.

100

• • • • •

We're two creeks between three lakes,
Flowing northeast to southwest,
Our contents slightly acidic,
Upstream dilation if compressed.

Hint #1:

pH of CSF is 7.3.

Hint #2:

These are diagonally oriented, bilateral structures (two creeks between three lakes).

The Interventricular Foramen of Monro

The interventricular foramen (of Monro) is a comparatively small channel that drains CSF from the lateral ventricles to the third ventricle (which are all much larger cavities of fluid, and thus "*two creeks between three lakes*"). Each of them is oriented diagonally and thus flow "*northeast to southwest.*"

The interventricular foramen of Monro carries CSF only, and CSF is more acidic than blood (CSF pH: 7.3 vs. plasma pH: 7.4, thus "*my contents slightly acidic*").

The foramen can get compressed by intracranial masses, which in turn can cause dilation of the corresponding lateral ventricle; hence "*Upstream dilation if compressed.*"

An important clinical application of this anatomy is found in colloid cysts; benign epithelial growths which are typically located in the third ventricle. In most patients, these cysts are asymptomatic, but in some cases they cause acute obstructive hydrocephalus by occluding the foramen of Monro.

A commonly used bedside neurosurgical procedure to alleviate hydrocephalus is a ventriculostomy. An ideally placed external ventriculostomy drain (EVD) is in the frontal horn of the lateral ventricle near the foramen of Monro.

This is shown in Figures A.1 and A.2.

101

● ● ● ● ●

After a split below the knee,
I descend shallow laterally,
Sensation dorsal foot I send,
If cut, ankle cannot outward bend.

Hint #1:

Do not dig too deep to find the answer.

Hint #2:

Goes by two names.

Superficial Fibular (Peroneal) Nerve

The superficial fibular nerve (SFN) is one of the two branches of the common fibular (peroneal) nerve. Its origin, the sciatic nerve, contains two fascicles – the common fibular and the tibial. These divide into two nerves in the distal posterior thigh. The common fibular nerve contains a medial fascicle that becomes the deep fibular nerve and a lateral fascicle that becomes the SFN at the fibular neck.

After branching from the common fibular nerve, the SFN travels inferiorly in the lateral leg, through the peroneus longus and peroneus brevis muscles, innervating them both, then into a groove between peroneus brevis and extensor digitorum longus. The SFN then pierces the deep fascia of the lateral leg and provides sensation to most of the skin of the lateral leg and dorsum of the foot.

While isolated injury to the SFN is uncommon, it is well described. Most commonly, the injury occurs in dancers and athletes who perform repeated foot inversion and plantarflexion. These motions can compress the SFN as it pierces the crural fascia of the lateral compartment and emerges into the subcutaneous fat. As a result, patients develop weakness of foot eversion (due to weakness of the peroneus longus and peroneus brevis) and diminished sensation on the dorsal surface of the foot. In SFN injury, strength of toe extension, foot inversion, and ankle dorsiflexion is unchanged. The sensation of the first webbed space (deep fibular nerve) and fifth toe (sural nerve) will be spared as well.

102

• • • • •

A branch of a more common nerve,
Dives deep in front it goes,
If lesion me can still evert,
But foot drop, numb 'tween toes.

Hint #1:

Cannot find the answer? Look in the tarsal tunnel.

Hint #2:

Almost completely motor. Almost.

Deep Fibular (Peroneal) Nerve

The deep fibular nerve (DFN) is one of the two branches of the common fibular nerve once it divides near the fibular neck. After branching, the deep fibular nerve travels inferomedially to the anterior leg slightly medial to the tibia, then descends to the leg adjacent to the anterior tibial artery, and deep to the extensor digitorum longus muscle. After passing anterior to the ankle joint, adjacent to the dorsalis pedis artery, it divides on the dorsum of the foot into its medial and lateral terminal branches.

The DFN has primarily motor function and innervates the anterior leg muscles that dorsiflex the ankle and extend the toes up toward the head. The medial terminal branch of the deep fibular nerve provides sensation to a small area of skin between the first and second toes.

Isolated injury of the DFN can occur at the ankle in the anterior tarsal tunnel syndrome. In this syndrome, the DFN is compressed between the extensor retinaculum (superficially) and the talus and navicular bones (deep). Unlike carpal or cubital tunnel syndromes, tarsal tunnel syndrome is rarely traumatic and usually involves a nerve tumor like a neuroma or ganglion cyst. Injury to the DFN causes weakness of dorsiflexion (tibialis anterior), weakness of toe extension (extensors digitorum and extensors hallucis), and some-times subtle weakness of foot eversion (peroneus tertius). Sensory loss is restricted to the first webbed space.

103

• • • • •

Branch from ascending artery,
Travel back to deep white,
If stop flow to the left limb,
Pure weakness to limbs right.

Hint #1:

One of the terminal branches of the internal carotid artery.

Hint #2:

Travels anterior to posterior.

Anterior Choroidal Artery

The anterior choroidal artery (AChA) is one of the terminal branches of the distal internal carotid artery after it enters the skull. It originates from the internal carotid artery just distal to the origin of the posterior communicating artery. The AChA measures only ~1 mm in diameter and travels posteriorly through the crural cistern, lateral to the optic tract and cerebral peduncle. It supplies blood to the optic radiations, posterior limb of the internal capsule, lateral thalamus, lateral geniculate nucleus, lateral portions of the cerebral peduncle, and portions of the mesial temporal lobe including the amygdala and the head of the hippocampus.

Clinically, occlusion of the AChA is typically caused by intracranial atheromatous disease. This occlusion will usually produce contralateral hemiparesis by infarcting the corticospinal tract as it descends through the posterior limb of the internal capsule. Less commonly, patients will suffer additional dysarthria (ischemia of the corticobulbar tract) and/or contralateral hemisensory loss (ischemia of the superior thalamic radiations). These latter symptoms tend to recover after injury.

Other peculiarities of anterior choroidal artery occlusion syndrome are the two atypical hemianopias it can cause. Because the AChA shares responsibility for perfusion of the optic radiations, its occlusion can cause an incomplete or incongruous homonymous hemianopia, where the visual field deficits it causes in the right and left eye are not the same. Meanwhile, if the AChA occlusion causes hypoperfusion to part of the lateral geniculate nucleus, the patient can suffer homonymous hemianopia with sparing of a beak-shaped zone in the middle, aka quadruple sectoranopia.

This is shown in Figure A.3.

104

• • • • •

From ACA, I branch and trail,
To feed the head of Latin *tail*,
If both vessels clot in pair,
You will not move or talk or care.

Hint #1:

Latin for "tail" is *cauda*.

Hint #2:

One of the striate arteries.

Recurrent Artery of Heubner/Medial Striate Artery

First described by and named after the German pediatrician Johann Otto Leonhard Heubner in 1872, the medial striate artery or recurrent artery of Heubner is a branch of the anterior cerebral artery, most often originating near the division of the anterior communicating artery. Though very small in diameter (~0.8 mm), it supplies blood to multiple structures including the head of the caudate ("tail"), medial portion of the globus pallidus, anterior limb of the internal capsule, nucleus accumbens, and the basal nucleus of Meynert. While classically assumed to be a single vessel, up to 75% of patients have multiple (up to four) arteries of Heubner branching off the same anterior cerebral artery.

Occlusion of the recurrent artery of Heubner is typically caused by atheromatous disease and can produce ischemia at the head of the caudate. Motor symptoms like contralateral hemiparesis are well described but usually transient. Instead, the most prominent clinical feature of these infarcts is abulia (a triad of minimal spontaneous output, paucity of response, and long response latency). This is thought to be caused by damage to the caudato-nigro-thalamo-cortical circuits, which are responsible for thinking, planning, and other high-level functions. Paradoxically, some patients with a right caudate infarct develop restlessness and agitation, and some patients even wax and wane between abulic and agitated states.

105

• • • • •

Climbs the neck, but starts at four,
After two-way divide,
Into canal, and through sinus,
Most half-brain blood supplied.

Hint #1:

Starts inferiorly dilated and decreases diameter as it ascends.

Hint #2:

Responsible for anterior cerebral circulation.

Internal Carotid Artery

The internal carotid artery (ICA) is a large blood vessel originating in the neck and terminating in the skull. The terminal branches of the common carotid artery, the internal and external carotid arteries (ECA), begin at the level of the C4 vertebra and ascend. Shortly after its division from the common carotid artery, the first 1-3 centimeters of the ICA are known as the carotid sinus, which serves as a baroreceptor to help regulate blood pressure.

There are several segments to the ICA, and anatomists and surgeons have divided it anatomically in many ways. One of the most common is the Bouthllier classification created in 1996, which divides the ICA into seven sections (C1–C7) as it completes a series of 90 degree turns through its course. After the common carotid artery terminally divides into the ECA and ICA, the ICA's first anatomical segment is the *cervical segment* (C1), which ascends to the neck within the carotid sheath (with the jugular vein, vagus nerve, and sympathetic plexus), slightly posterolaterally to the ECA, eventually entering the skull via the carotid canal.

The *petrous segment* (C2) then turns 90 degrees anteromedially and travels through the petrous temporal bone of the carotid canal. After exiting the carotid canal, the ICA immediately turns 90 degrees superiorly again as the *lacerum segment* (C3), ascending to the petrolingual ligament, at which point it again turns 90 degrees anteriorly, entering the cavernous sinus as the *cavernous segment* (C4) of the ICA. After traveling anteriorly through and exiting the cavernous sinus, the ICA again turns 90 degrees and travels superiorly for a very short segment called the *clinoid segment* (C5) as it passes through the dura, before it again turns 90 degrees posteriorly as the *ophthalmic/supraclinoid segment* (C6), passing medial to the clinoid processes. Finally, the ICA again turns superiorly, ending its journey at the *terminal/ communicating segment* (C7), where it then divides into the anterior cerebral artery, middle cerebral artery, anterior choroidal artery, and posterior communicating artery.

In vascular neurology, ICAs are often found to have atheromatous disease and less commonly can suffer dissections. Extracranial ICA atheromatous disease typically forms just distal to the common carotid bifurcation, while intracranial ICA atheromatous disease is more common in the cavernous (C4) and supraclinoid (C6) segments. In general, ICA atheromatous disease presents two major stroke risk factors. First, as the ICA becomes more stenotic, the flow rate increases through the area of stenosis, increasing the chance of flicking a piece of atheromatous plaque into one of the intracranial vessels. Second, ICA stenosis can eventually become flow-limiting, such that in the case of systemic hypotension, the addition of flow-limiting stenosis can cause cerebral hemispheric hypoperfusion. Because of collateral circulation, these infarcts occur in watershed zones, typically causing proximal weakness at the contralateral shoulder and hip girdles. Because patients can move their distal extremities but are unable to move their arms or thighs, this clinical picture is nicknamed "Person in a Barrel Syndrome."

This is shown in Figure A.3.

.

106

• • • • •

Branch from near top of basilar,
Passing under third nerve,
Through ambient cistern, if clot,
Cerebellar stroke observe.

Hint #1:

Just below the PCA.

Hint #2:

Look to the top of the cerebellum.

Superior Cerebellar Artery

The superior cerebellar artery (SCA) is the second most superior branch from the basilar artery, branching bilaterally just inferior to the origin of the posterior cerebral artery. From its origin at the basilar artery, the SCA first travels laterally, anterior to the pons in the prepontine cistern, passing inferior to the oculomotor nerve that separates it from the posterior cerebral artery. The SCA then curves posteriorly and travels through the ambient cistern, passing just inferior to the trochlear nerve before traveling posteromedially to the lateral edges of the quadrigeminal cistern, coming in close proximity to the contralateral SCA. The SCA perfuses the inferior portions of the midbrain and lateral pons, the superior and middle cerebellar peduncles, the superior portions of the cerebellar vermis, and the dentate nucleus.

Occlusion of the SCA can either be caused by direct infiltration of an embolus into the SCA or by a larger embolus getting caught at the rostral end of the basilar artery, producing "top of the basilar syndrome." Diminished blood flow to the superior cerebellum can cause both gait and limb ataxia. Because the SCA supplies the dentate nucleus and superior cerebellar peduncle, an SCA-territory infarct can also result in ipsilateral tremor or even chorea.

This is shown in Figure A.3.

107

• • • • •

A branching from L2 and 3,
'Neath ligament it goes,
And medial to spine of hip,
If pinched, thigh pain one knows.

Hint #1:

Contrary to one of its names, it is not a branch of the femoral nerve.

Hint #2:

Keep this nerve happy by loosening your belt.

237

Lateral Cutaneous Nerve of the Thigh

Though it is also called the *lateral femoral cutaneous nerve*, the name lateral cutaneous nerve of the thigh has become preferred, given the nerve is not a branch of the femoral nerve, but rather a nerve originating in the lumbar plexus. From the posterior division of the L2 and L3 nerve roots, the lateral cutaneous nerve of the thigh travels inferolaterally, just under the iliolumbar ligament and lateral to the psoas major muscle, then courses through the iliac fossa lying on the anterior surface of the iliacus muscles, ultimately passing deep to the inguinal ligament and just medial to the anterior superior iliac spine. After passing under the inguinal ligament, it travels over the sartorius muscle before dividing into anterior and posterior branches. The anterior branch contains mainly L3 fibers and provides sensation to the anterolateral thigh above the knee, while the posterior branch possesses mainly fibers from L2 and provides sensation to the lateral thigh over the iliotibial band.

Impingement of the lateral cutaneous nerve of the thigh typically occurs at the inguinal ligament. As expected, this can cause numbness or dysesthesias of the lateral thigh, commonly known as "lateral femoral cutaneous syndrome" or "meralgia paresthetica." This sensory mononeuropathy is classically seen in patients who wear tight belts or pants, have gained weight recently, and/or have diabetes mellitus. Most cases are cured with conservative measures alone, which include weight loss and/or wearing looser-fitting clothing. Pharmacologic or procedural therapy is rarely needed.

108

• • • • •

A nerve with roots from 5 and 6,
Gives upper arm sensate,
Abducts most range and is injured,
If shoulder dislocate.

Hint #1:

Find one of its branches in the glenohumeral joint.

Hint #2:

One of five terminal plexus branches.

Axillary Nerve

The axillary nerve is one of the terminal branches of the brachial plexus. A branch of the posterior cord, the axillary nerve is composed of nerve fibers originating from C5 and C6 roots. From the brachial plexus, the axillary nerve courses inferolaterally anterior to the subscapularis muscle, just posterior to the axillary artery and beneath the glenohumeral joint, ultimately passing through the quadrangular space. Here, it terminally divides into its anterior and posterior branches.

The anterior branch wraps around the surgical neck of the humerus and provides motor innervation to the anterior deltoid muscle, which functions as the primary abductor of the arm at the shoulder. The anterior branch also terminates in small sensory branches that project to the skin of the anterior and lateral shoulder.

The posterior branch provides motor innervation to the posterior deltoid and teres minor muscles, as well as sensory innervation to the skin of the upper lateral arm through its terminal superior lateral brachial cutaneous branch.

The axillary nerve also has articular branches that provide sensation to the glenohumeral joint.

Injury to the axillary nerve typically occurs in conjunction with other nerve injuries, such as brachial plexopathy. When isolated axillary mononeuropathy occurs, it is typically due to a stretching injury within the quadrangular space, most often caused by anterior shoulder dislocation. Those with axillary nerve injury will have weakness of shoulder abduction between 30 and 90 degrees (function of deltoid muscle). External rotation of the arm will also be affected but may be difficult to detect on examination, given infraspinatus is also heavily involved in external rotation. Axillary neuropathy in concert with other nerve injuries may be seen in disorders like traumatic brachial plexopathy or neuralgic amyotrophy (Parsonage–Turner syndrome).

109

• • • • •

Supplied by roots L2, 3, 4,
Like-named canal enclosed,
Senses the skin of inner thigh,
Adduct the thigh it shows.

Hint #1:

Pierces a like-named muscle in the thigh.

Hint #2:

Also provides articular branches to the knee.

Obturator Nerve

The obturator nerve originates from the anterior rami of L2–L4 through the lumbar plexus. It travels inferiorly into the pelvis medial to both the femoral nerve and psoas major muscle, exiting the pelvis just inferior to the pubic bone along with the obturator vessels via the obturator canal. The obturator nerve splits into two terminal branches, the anterior and posterior divisions of the obturator nerve, around the adductor brevis muscle as it exits the pelvis into the thigh. The anterior division ultimately provides motor innervation to adductor longus, adductor brevis, and gracilis, as well as sensory innervation to portions of the inner thigh and the hip joint. The posterior division pierces and innervates obturator externa on its way out of the pelvis, as well as providing motor innervation to adductor magnus and articular/genicular branches of the medial knee.

The rare incidence of isolated obturator neuropathy is best studied in sports injuries. During exercise, the nerve is thought to be impinged on its course through the adductor compartment due to fascial thickening and may be worsened by further entrapment from vessels of the medial circumflex femoral artery. The most typical complaint in cases of obturator neuropathy is paresthesia or pain of the groin and inner thigh. Over time, the nerve injury can also cause denervation of adductors longus and magnus, leading to clinical weakness of thigh adduction. Obturator neuropathies are typically treated conservatively, with only the most debilitating cases requiring nerve blocks or surgical intervention.

110

• • • • •

This branches from the tenth in neck,
Loops vessels, then ascend,
To innervate most larynx,
If cut, smooth voice will end.

Hint #1:

Innervates the vocal cords.

Hint #2:

Look between the trachea and esophagus.

Recurrent Laryngeal Nerve

The recurrent laryngeal nerve originates from the vagus nerve near the cervicothoracic junction. The right recurrent laryngeal nerve branches from the vagus nerve at or slightly above the level of the right clavicle, after the vagus nerve passes over the right subclavian artery. After branching from the vagus nerve, the right recurrent laryngeal nerve loops posteriorly under the right subclavian artery and then ascends back up the neck, traveling in the right tracheo-esophageal groove on its ascent. The left recurrent laryngeal nerve has a slightly different path as it originates from the vagus nerve just below the level of the left clavicle and loops under the aortic arch before ascending in the left tracheoesophageal groove. Both nerves then ascend deep to the thyroid gland to eventually innervate the intrinsic muscles of the larynx (with the exception of the cricothyroid muscle) and provide sensation to the trachea and larynx below the level of the vocal cords.

Injury and subsequent dysfunction of the recurrent laryngeal nerve can occur from insult anywhere on its extensive pathway throughout the neck and chest. The most common cause is accidental transection during head or neck surgery, most classically during a thyroidectomy. Other causes of recurrent laryngeal neuropathy include compressive effect from tumors or iatrogenic injury during endotracheal intubation. These injuries, which result in unilateral vocal cord paralysis, typically result in a hoarse monotone voice. Monotony specifically occurs because of denervation of the thyroarytenoid muscle, which modulates pitch. Treatment can range from conservative measures such as speech therapy to surgical interventions such as reinnervation or tumor resection.

111

• • • • •

I house cell bodies of the nerve,
With signals to the cord,
Sensation passes right by me,
But no synapse is stored.

Hint #1:

Pea-sized or smaller.

Hint #2:

Consider the term "pseudounipolar."

Dorsal Root Ganglion

The dorsal root ganglion (DRG) is a collection of sensory nerve cell bodies located outside the spinal cord at each vertebral level. These are pseudounipolar neurons with an axon that bifurcates and travels in two directions. The peripheral part of this axon receives distal sensory information from elsewhere in the body and projects it proximally to the DRG, which is located in or just distal to the intervertebral foramen at each spinal level, corresponding to the dermatome of that sensory area. From there, the central axon transmits the signal to the spinal cord. The function of the DRG is to transmit sensory information from the nociceptors, chemoreceptors, proprioceptors, and thermoreceptors of the peripheral nervous system to the central nervous system. Unlike most other ganglia, the DRGs do not contain any synapses, only the cell bodies of the pseudounipolar neurons.

In evaluation of patients with suspected peripheral neuropathy versus radiculopathy, SNAPs on electrodiagnostic testing play a significant role. If the lesion is at the level of the DRG (ganglionopathy or neuronopathy) or distal to it (peripheral neuropathy), SNAPs will be abnormal. If the lesion is proximal to the DRG (radiculopathy), SNAPs will be normal.

Dorsal root ganglionopathy or neuronopathy is a non-length-dependent pure sensory neuropathy (not necessarily affecting the longest peripheral nerves first), which presents with sensory symptoms, ataxia, and hyporeflexia without weakness. Major differential diagnoses include infectious (HIV, VZV, leprosy), autoimmune (Sjogren's syndrome), paraneoplastic (small-cell lung carcinoma), and toxic (carboplatins, vitamin B6) etiologies.

112

• • • • •

All sensory and no motor,
From the femoral it split,
Travels inner thigh to arch,
Med' calf numb if cut it.

Hint #1:

A superficial nerve.

Hint #2:

Runs with a like-named vein.

Saphenous Nerve

The saphenous nerve is a terminal branch of the femoral nerve and functions as a pure sensory nerve, providing cutaneous innervation to the medial knee, leg, ankle, and foot. It originates from the posterior division of the femoral nerve in the femoral triangle, lateral to the femoral artery and vein. Exiting the femoral triangle, it enters the adductor canal and travels inferiorly with the femoral artery and vein. The saphenous nerve then exits the canal, just superior to the knee, travels superficially, and gives off its first branch, the infrapatellar branch, which travels through the inferior portion of the sartorius muscle and provides sensory innervation to the skin of the inferomedial knee. The saphenous nerve continues inferiorly, traveling superficially down the medial leg with the saphenous vein and passing anterior to the medial malleolus at the ankle and ultimately to the great toe. Interestingly, its name is derived from the Greek word *safaina,* meaning "evident," thought to have been named because of its superficial position under the skin.

Chronic irritation of the saphenous nerve as it passes through the adductor canal on its path toward the knee produces a syndrome known as saphenous neuritis or "Gonalgia Paresthetica." As a result of entrapment with or without direct trauma, saphenous nerve inflammation leads to pain of the anteromedial aspect of the knee and sometimes radiates down the nerve's typical course. It can often be diagnosed on physical exam of a person with knee pain, where direct palpation of the saphenous nerve along its pathway reproduces the pain (similar knee pain syndromes like a medial meniscus tear typically lack this key finding). As with most entrapment neuropathies, treatment often involves an escalating approach from conservative to pharmacologic to procedural care. Many patients respond well to physical therapy and stretching exercises with coadministration of nonsteroidal anti-inflammatory drugs (NSAIDs), while some may require local nerve injections or, in the most extreme states (like the formation of a saphenous neuroma), surgical decompression or neurectomy.

113

• • • • •

With origins from only roots,
5 through 7, called *nerve of Bell*,
With motor only to one place,
Winged scap' if not work well.

Hint #1:

Cannot think of it, take a deep breath in.

Hint#2:

Find the muscle it innervates under your armpit.

Long Thoracic Nerve

The long thoracic nerve is a pure motor nerve with fibers from nerve roots C5 through C7. It descends posteriorly through the roots of the brachial plexus, emerging between the middle and posterior scalene muscles, then travels along the outer lateral thoracic wall along the midaxillary line just superficial to the serratus anterior muscles. The sole function of the long thoracic nerve is to provide motor innervation to the serratus anterior muscle. This muscle functions to protract, upwardly rotate, and stabilize the scapula. It also serves as an accessory muscle of respiration during inhalation. It is sometimes referred to as the "Boxer's Muscle" because it protracts the scapula when throwing a punch.

Long thoracic nerve injury is a classic disorder taught in medical school because of the unique "winged scapula" appearance it produces. This injury is often produced as a surgical complication of thoracic intervention, including anything from bedside procedures like chest tube placement to major surgeries like radical mastectomies. In these cases, the nerve is typically transected on its pathway along the eighth and ninth ribs, where it runs mostly unprotected. When the long thoracic nerve is injured and the serratus anterior weakened, the tonic balance between the rhomboid muscles and the serratus anterior is disrupted. As a result, the scapula is pulled medially and projects posteriorly, giving an "angel wing" appearance. It's for this reason that medical students are often taught the mnemonic "C5, 6, and 7 keep the wings from going to heaven."

114

• • • • •

Roots 6 through 8, motor only,
From armpit down back send,
Branch of the cord posterior,
To arm adduct/extend.

Hint #1:

Innervates the "broadest" muscle in Latin.

Hint #2:

Find passing just under teres major.

Thoracodorsal Nerve

The thoracodorsal nerve, also known as the long subscapular or middle subscapular nerve, is a pure motor nerve originating from spinal nerve roots C6 through C8, exiting the brachial plexus after branching off the posterior cord. It then travels inferiorly down the posterior axillary wall, passing under the inferior edge of the teres major muscle and ultimately innervating the latissimus dorsi muscle (name meaning "broadest back" in Latin), aiding in adduction, medial rotation, and extension of the arm at the elbow.

The thoracodorsal nerve has a variety of clinical correlations, the most common of which includes accidental injury during procedures leading to difficulty with shoulder adduction and overhead pull-down (from weakness of the latissimus dorsi). However, the nerve and its muscle are sometimes intentionally sacrificed to give function to other areas of the body. For patients who have suffered devastating injuries like transections to the facial or musculocutaneous nerves, the lateral branch of the thoracodorsal nerve can be used for nerve graft reconstructive surgery. Likewise, the latissimus dorsi itself can be used as a flap for breast reconstructive surgery.

115

• • • • •

From top trunk C6 and 5,
Gives shoulder joint sensate,
Abducts the arm, the first 15,
Motor to both spinate.

Hint #1:

Passes through a like-named notch.

Hint #2:

Does not go to the arm or hand.

Suprascapular Nerve

The suprascapular nerve is supplied by the C5 and C6 nerve roots and is the only nerve that originates from the upper trunk of the brachial plexus. After branching from the brachial plexus, it travels laterally with the suprascapular artery and vein, deep to the trapezius muscle, and passes through the suprascapular notch, bordered above by the suprascapular ligament, into the supraspinatus fossa. There, it innervates the supraspinatus muscle before continuing into the infraspinous fossa via the spinoglenoid notch and innervating the infraspinatus muscle. On its way, it provides sensory articular branches to both the glenohumoral and acromioclavicular joints. The supraspinatus is an arm abductor at the shoulder joint for the first 15 degrees of abduction. The infraspinatus helps with external rotation and extension of the arm at the shoulder joint.

Suprascapular nerve injury can cause weakness of either the infraspinatus alone or both the infraspinatus and supraspinatus together, depending on where along its course the nerve is injured. Iatrogenic injuries (such as those incurred during rotator cuff repairs) typically injure the proximal suprascapular nerve and thus affect both muscles. Similarly, a compression injury to the nerve at the suprascapular notch occurs proximal to its bifurcation, resulting in weakness of both muscles. The patients thus have the inability to initiate arm abduction (supraspinatus) or perform external rotation (infraspinatus). By contrast, if the nerve is injured after the bifurcation at the level of spinoglenoid notch, the patient will only have weakness of arm external rotation due to isolated denervation of the infraspinatus.

116

• • • • •

I'm lateral, just sensory,
Nine-ninety round my axis,
With three canals, I feel the wave,
As it wanes and waxes.

Hint #1:

Located in the petrous portions of the bilateral temporal bones.

Hint #2:

Fluid-filled.

Cochlea

The cochlea is a spiral-shaped portion of the inner ear osseous labyrinth, located in the petrous portion of the temporal bone. It contains the cochlear duct, a portion of the membranous labyrinth that senses and processes sound. Its spiraled contour rotates a total of approximately two and three-quarter turns (990 degrees) around a central portion of bone called the modiolus. This spiraled tube is subdivided into three fluid-filled tubes by two membranes, the spiral lamina/basilar membrane and Reissner's membrane. Two of the tubes, the scala tympani and scala vestibuli, considered the lower and upper sections respectively, are filled with perilymph. The middle tube is called the scala media (or cochlear duct), which is filled with endolymph and contains the organ of Corti, the primary sensory organ for hearing.

Sound waves hitting the tympanic membrane cause movement of the middle ear ossicles – malleus, incus, and stapes. These vibrations are then transmitted to the cochlea via the oval window. The vibrations cause the basilar membrane to move against Reissner's membrane, which activates the hair cells. This activates the nerve fibers whose cell bodies are in the spiral ganglion. These neurons in turn project to the cochlear nuclei in the medulla via the cochlear nerve.

Pathologies affecting the cochlea and/or the cochlear nerve fibers produce ipsilateral sensorineural hearing loss. The auditory pathway after the cochlear nuclei in the brainstem ascends bilaterally. Hence, pathologies affecting this pathway after the level of the cochlear nuclei are unlikely to produce hearing loss.

117

• • • • •

From Latin *net* and both front set,
Absorb photons, change then,
That energy into signal,
Up is down, and out in.

Hint #1:

Extremely metabolically active.

Hint #2:

The word for "net" in Latin is *rete*.

Retina

The retina is a thin layer of sensory neurons lining the posterior-most two-thirds of the inner eyeball. The retina functions specifically to sense light energy in the form of photons coming in from the environment through the lens and processing it into neuronal signals. The retina and the optic nerve are derived from the diencephalon.

The retina has three major cellular layers (and seven minor ones). From inside (toward the vitreous) out (toward the orbital apex), these major layers are the ganglion cell layer, bipolar cell layer, and photoreceptor cell layer.

The reason for this paradoxical inside-out organization of the retina is not known. The light goes through all the layers and is then detected by the photoreceptors in the back. The two major types of photoreceptors are rods (sensitive to light) and cones (used for color detection). Interestingly, the photoreceptors do not get depolarized by the photons; instead, they get hyperpolarized. In addition to these main cellular layers, there are also amacrine cells and horizontal cells, which produce lateral excitation or inhibition at nearby ganglion cells and bipolar cells, respectively. This allows bipolar neurons and ganglion cells to have a center-surround receptive field.

Cells that are excited when light falls in the center of their receptive field are referred to as on-center cells. Off-center cells are those that get inhibited by light in their receptive field.

The highest concentration of the cones is at the fovea. At the fovea, the retinal ganglion cell layer is the thinnest, and the cells are moved away to allow for light to reach the photoreceptors directly.

Overall, the retinal ganglion cells are of broadly two types – parasol cells and midget cells. Parasol cells are larger and have larger receptive fields. These are sensitive to movements and gross features of stimuli. These cells project to the magnocellular layer of the lateral geniculate nucleus. The midget cells are smaller with smaller receptive fields and are sensitive to finer details of the stimuli, including color. These cells project to the parvocellular layer of the lateral geniculate nucleus.

Because the retina is highly metabolically active and its tissue requires more oxygen than any other tissue in the body, it gets dual blood supply, with the inner layers supplied by the central retinal artery and the outer layers supplied by the choriocapillaris, which is fed by the posterior ciliary artery. This large net-like area of blood supply to the posterior eye is thought to have inspired its name, derived from the Latin word *rete* meaning "net."

118

• • • • •

Three loops that are not quite complete,
With Latin *flask* at end,
We're filled with liquid that will move,
When turn your head or bend.

Hint #1:

"Flask" in Latin is *ampulla*.

Hint #2:

Contain uniquely functioning hair cells.

Semicircular Canals

The semicircular canals are specialized endolymph-filled structures in the inner ear that help regulate the position sense of the head and overall balance. Like the cochlea, the semicircular canals are bony labyrinths housed in the petrous portion of the temporal bone. Each inner ear has three semicircular canals on each side. Named the lateral, superior, and posterior canals, each is oriented at an approximately 90-degree angle from the other two, and each contoured to make approximately two-thirds of a full circle.

At the base of each of the semicircular canals is a bulbous dilation called the ampulla. The ampulla houses specialized sensory structures called the crista ampullaris. The crista contains hair cells that project into a gelatinous structure called the cupula. Angular acceleration of the head in a certain direction causes the endolymph to flow, which distorts the cupula and moves the hair cells. The activation of hair cells activates the vestibular nerve fibers, whose cell bodies are located in the superior and inferior vestibular ganglia. These neurons then project to the vestibular nuclei in the medulla.

Approximately one-fifth of physician visits for vertigo turn out to be caused by a benign but vexing condition of the semicircular canals called benign paroxysmal positional vertigo (BPPV). BPPV occurs when otoconia from the otolith organs get dislodged and enter one of the semicircular canals, causing movement of the endolymph in the absence of actual head movement. This causes slow eye movement in the plane of the semicircular canal with a corrective fast phase (nystagmus) and a perception of motion (vertigo).

119

• • • • •

On medial surface cortex,
I carve a curv'ed path,
In occiput I separate,
Vision cortex in half.

Hint #1:

Derived from Latin "spur-shaped."

Hint #2:

I end posteriorly at the pole.

Calcarine Sulcus

Also called the calcarine fissure, the calcarine sulcus is located on the medial surface of both occipital lobes, dividing the primary visual cortex into upper and lower portions, namely, the cuneus and the lingual gyrus, respectively. Deriving its name from the Latin word *calcar* meaning "spur," it travels a slightly curved path stretching anteriorly to the parieto-occipital fissure and posteriorly to the occipital pole. Its blood supply comes from a branch of the posterior cerebral artery called the calcarine artery.

The primary visual cortex (V1) aka Brodmann area 17 is located in the banks of the calcarine sulcus. The superior bank receives superior optic radiations from the lateral geniculate nucleus, and the inferior bank receives the inferior optic radiations (Meyer's loop). Visual information has a retinotopic organization all the way from the retina to the peri-calcarine cortex. Hence, a lesion of the superior bank of the calcarine sulcus will produce a contralateral inferior quadrantopia, and vice versa.

The information arriving from the lateral geniculate nucleus ends in layer 4 of V1. Specifically, information from magnocellular layer ends in layer $4C\beta$ and that from the parvocellular layer ends in layer $4C\alpha$. Interestingly, the information from each eye is not organized in layers but as ocular dominance columns in V1.

Because of extensive afferents reaching V1, layer 4 contains thick myelinated axon collaterals, referred to as Stria of Gennari. Hence, the name striate cortex for the primary visual cortex.

This is shown in Figure A.1.

120

• • • • •

Placed in rear, a Latin *cone*,
Top portion of a lobe,
Interpreting the other side,
And bottom field of globe.

Hint #1:

Important area for visual fields.

Hint #2:

Contains radiations.

Cuneus

The cuneus is the superior portion of the occipital lobe, positioned above the calcarine sulcus. It is bordered anteriorly by the parieto-occipital fissure and posteriorly comes to a point at the occipital pole. It receives dual blood supply from the calcarine artery and the parieto-occipital artery. The cuneus functions as a part of the primary visual cortex, as its inferior border (the superior bank of the calcarine sulcus) houses synapsing fibers from the superior portion of the optic radiations. This part of the primary visual cortex analyzes visual information from the contralateral inferior quadrant of both visual fields and is implicated in basic visual processing.

Isolated cuneus injury is rather rare and characterized by a contralateral inferior homonymous quadrantanopia and clinically appears very similar to an injury of the superior optic radiations. However, what the cuneus lacks in common neurologic injury it makes up for with intrigue in modern functional neuroanatomy studies. Specifically, atrophy of the cuneus has been found in association with a variety of disorders accompanied by visual hallucinations including schizophrenia, postencephalitic psychosis, and Lewy body dementia syndrome.

This is shown in Figure A.1.

121

• • • • •

Supplied by blood from PCA,
Greek – *glossus*, is so named,
Cannot see top other side field,
If one side tissue maimed.

Hint #1:

Injury can cause a quadrantanopia.

Hint #2:

Named after its tongue-like shape.

Lingual Gyrus

The lingual gyrus is the portion of medial occipital lobe just inferior to the calcarine sulcus, and so named because of its *tongue-like* shape. The lingual gyrus functions as a part of the primary visual cortex, as its superior border (the inferior bank of the calcarine sulcus) houses synapsing fibers from Meyer's loop. This part of the primary visual cortex analyzes visual information from the contralateral superior quadrant of both visual fields.

The lingual gyrus is also known as the medial occipitotemporal gyrus, as it continues into the base of the temporal lobe. The basal temporal region is anatomically indistinct from the occipital lobe. Anteriorly, the lingual gyrus transforms into the parahippocampal gyrus.

Isolated injury to the unilateral lingual gyrus results in contralateral superior homonymous quadrantanopia, though this is more rarely seen than the identically presenting injury to Meyer's Loop. However, dysfunction of the lingual gyrus and the nearby fusiform gyrus is also associated with several additional unique and fascinating visual processing disorders such as dyschromatopsia and prosopagnosia. Dyschromatopsia is a condition where patients lose the ability to process colors, resulting in either the perception of black-and-white vision or even discolored vision (like seeing all hard colors as brown and soft colors as gray). Prosopagnosia refers to poor facial recognition, including difficulty assessing features such as emotion, ethnicity, or age.

This is shown in Figure A.1.

122

· · · · ·

Once fluid's in the fourth,
There's only three ways that it leaves,
One that goes straight down,
And two laterally placed sleeves.

Hint #1:

A conduit for CSF to reach the subarachnoid space.

Hint #2:

Popularly known by and named after the anatomists who first described them.

Foramina of Luschka and Magendie

The foramina of Luschka and Magendie are three pathways through which CSF flows out of the fourth ventricle. The foramina of Luschka, also called the lateral apertures of Luschka, and named after the nineteenth-century German anatomist Hubert von Luschka, are left and right lateral pathways that allow CSF to flow out of the fourth ventricle into the subarachnoid space via the cerebellopontine cisterns. These foramina lie just posterior to the origin of the glossopharyngeal nerve (CN IX).

The foramen of Magendie, also often called the median aperture, is a midline pathway extending from the lower part of the roof of the fourth ventricle that allows CSF to flow posteriorly to the CSF space via the cisterna magna and quadrigeminal cistern.

One well-recognized but underdiagnosed disorder of the foramina is the Blake pouch cyst, which lies on the Dandy–Walker continuum. The Blake pouch is a normal neuroembryologic structure that forms and then perforates in the first trimester of gestation to form the foramen of Magendie. If the Blake pouch fails to perforate, CSF pressure can build and distend the inferior medullary velum, causing the "cystic" appearance on imaging. The foramina of Luschka form soon after and are typically able to compensate with increased CSF flow. If the compensation is inadequate, patients with Blake pouch cysts may develop increased intracranial pressure and subsequent headaches in childhood or young adulthood.

123

• • • • •

Two Latin words for *see-through wall*,
I sit midsagittal,
I separate the two great lakes,
And in variant get full.

Hint #1:

Made of two adhered membranes.

Hint #2:

Composed largely of ependymal cells.

Septum Pellucidum

The septum pellucidum is a thin membrane located midline in the brain between the left and right anterior horns of the lateral ventricles. Made up of only two adhered membranes, the septum pellucidum gets its name from the Latin meaning "transparent wall" and is comprised mainly of ependymal cells and pia mater.

The septum pellucidum is usually collapsed and vertically oriented. In cases of mass lesions in either hemisphere, its appearance gets distorted as viewed on neuroimaging in the coronal plane. Tumors and acute infarctions (among many other disorders) cause edema of the brain parenchyma and thus displace other nearby tissues. Displacement of brain tissue from one cavity to another is known as "herniation," with subfalcine herniation being the most common. The degree of subfalcine herniation is quantified with the measurement of "midline shift," or the degree to which the midline structures (namely, the septum pellucidum) have been displaced from normal position.

The septum pellucidum is also prone to a variety of developmental variants. Embryologically, the septum pellucidum begins as two separate septa running perpendicular that begin to adhere to each other. The caudal (posterior) aspects of these septa fuse at late-second and early-third gestational trimesters. Failure to do so results in a common anatomic variation known as "cavum vergae." Meanwhile, the remainder of the septal segments fuses in childhood and young adulthood; failure to do so results in "cavum septum pellucidum" (less commonly termed "cavum septi pellucidi"). Because of the sequential nature of fusion of the septum, essentially every patient with cavum vergae will also have cavum septi pellucidi. The two are seen in approximately 15% of the adult population.

124

• • • • •

The area of gray in cord,
Descending signals flow,
Upper and lower synapse here,
Injured, motor will not go.

Hint #1:

Houses synapses between central and peripheral nerves.

Hint #2:

Found only in the spinal cord.

Anterior Horn of the Spinal Cord

The anterior horn of the spinal cord, also known as the ventral horn, is an area of gray matter positioned anterolateral to the central canal of the spinal cord. It contains the alpha motor neurons, which give rise to motor axons via the ventral root for the mixed peripheral nerves. These cells are located at every spinal level, though are more highly concentrated in the cervical and lumbar regions, given their projection to the extremities through the brachial and lumbar plexi, respectively.

Diseases of the anterior horn cells are characterized by typical findings of lower motor neuron dysfunction, which include flaccid weakness, hyporeflexia, denervation atrophy, and electromyographic study findings of neurogenic motor units and fibrillations. These notably occur in the absence of sensory symptoms, which helps distinguish anterior horn disorders from progressive polyneuropathies.

Some important diseases selectively affecting alpha motor neurons in the anterior horn cells include viral myelitis syndromes such as polio and West Nile virus along with neurodegenerative disorders such as spinal muscular atrophy and amyotrophic lateral sclerosis (ALS).

125

• • • • •

Midline and above cerebel',
Five-point of venous flow,
From the Latin *twist* or *wine press*,
If clot, head pain will know.

Hint #1:

Found just below the occipital pole.

Hint #2:

Its walls are made of dura mater.

Torcula/Confluence of Sinuses

The confluence of sinuses, also known as the torcula, is the most posterior area in the cerebral venous sinuses. It is the area in which the superior sagittal sinus (superior), straight sinus (anterior), and occipital sinus (inferior) meet and flow into both transverse sinuses (lateral). It is positioned superficially between the occipital lobe and the cerebellum, and like all cerebral venous sinuses, its walls are made of the dura mater.

Interestingly, one of its many names, "torcula," is short for torcula herophili, and so named after the fourth-century Greek surgeon and anatomist Herophilus. Though after being translated into many languages, the meaning is not completely certain, many believe its name is derived from either the Latin word *torculo* meaning "press" (as in for wine or olives), or from the Latin *torqueo* meaning "twist."

Thrombotic occlusions on the venous sinuses are emergent conditions, requiring prompt recognition and treatment. Often caused by states of hypercoagulability (such as pregnancy or malignancy), a thrombus forms within the venous system and causes backup of pressure into the proximal veins. This results in an obstructive communicating hydrocephalus with typical features of elevated intracranial pressure headache, including visual changes and worsening symptoms with lying down/Valsalva maneuver.

Focal neurologic deficits are incurred when the backpressure causes one of the veins to perforate and results in intracranial hemorrhage. The mainstay of pharmacologic treatment for this is therapeutic systemic anticoagulation, often with a continuous heparin infusion. Procedural intervention with mechanical thrombectomy can also be considered.

This is shown in Figure A.4.

126

• • • • •

Midline venous pathway which,
Brings blood from front to straight,
It drains brainstem and cerebel',
To earn its nickname "great."

Hint #1:

Despite its name, it is often very small.

Hint #2:

Fed by the basal veins of Rosenthal and the internal cerebral veins.

Great Cerebral Vein (of Galen)

The great cerebral vein, also called the great cerebral vein of Galen, is a deep midline cerebral venous sinus. It drains venous blood from bilateral basal veins of Rosenthal and the internal cerebral veins. It further drains superoposteriorly, eventually draining into the straight sinus. Though it only travels a relatively short distance (0.5–3 cm), it plays a critical role in helping venous drain of the blood from the brainstem, deep cerebral structures, and several posterior cranial fossa structures.

Vein of Galen aneurysmal malformations (VGAMs), while rare, represent a serious and dramatic disorder of infancy and early childhood. During the first trimester of gestational development, the median prosencephalic vein (embryologic precursor to the vein of Galen) forms as an arteriovenous shunt to drain the choroidal arteries that perfuse the early developing brain parenchyma. These structures then typically regress as cerebral perfusion converts to the major intracranial vessels.

Aberrant persistence of the arteriovenous malformation to the median prosencephalic vein leads to dilation and ballooning of the vein, producing a VGAM. This often occurs comorbid with high-output cardiac failure and has a grim natural history, as patients with this disorder have historically suffered from intracranial hemorrhages and fatal heart failure with hydrops fetalis.

127

• • • • •

From sensory afferent nerves,
We branch, go one or few,
Segments up or down through white,
Then synapse gray into.

Hint #1:

A posterolateral tract.

Hint #2:

Branches from first-order neurons.

Lissauer's Tract

Lissauer's tract, also known as the posterolateral tract of Lissauer, is a white matter tract that carries somatosensory fibers within the spinal cord.

The sensory fibers enter the posterior funiculus of the spinal cord via the dorsal root. The fibers that carry pain and temperature sensations are small in diameter and unmyelinated. As soon as these fibers enter the spinal cord, they synapse with the gray matter in the dorsal horn known as substantia gelatinosa, specifically laminae I and V. After synapsing, these axons cross over to the other side and ascend as the spinothalamic tract. However, some axon collaterals do not synapse and ascend or descend a few spinal segments as Lissauer's tract before synapsing with the dorsal horn gray matter.

One of the clinical pearls to remember during examination of a suspected myelopathy patient is that the actual lesion might be at a higher level than indicated by sensory examination. This is partially because of the existence of Lissauer's tract, as the axon collaterals of the sensory fibers can ascend several segments higher before decussating via the anterior commissure. This is particularly notable in the case of hemicord injury, also known as Brown–Sequard syndrome, where the patient's contralateral loss of pain and temperature is several spinal levels lower than the ipsilateral loss of motor function and vibratory/positional sense.

128

• • • • •

Adjacent to substance nigra,
Within the midbrain seen,
Reward, addict, and motivate,
By way of dopamine.

Ventral Tegmental Area

The ventral tegmental area (VTA) is a midline collection of dopaminergic neurons in the midbrain located between the left and right substantia nigra. There are three main dopaminergic pathways in the brain – nigrostriatal, mesolimbic, and mesocortical. The VTA is the major source of dopamine for the mesolimbic and mesocortical pathways.

As opposed to the nigrostriatal pathway, which is mainly concerned with motor control, the mesolimbic and mesocortical pathways are mainly concerned with reward, addiction, and behavioral modulation of movement.

The axons of the dopaminergic neurons terminate on the dendritic spines of medium spiny neurons in the striatum. Those from the substantia nigra pars compacta (nigrostriatal pathway) terminate on the dorsal striatum (caudate and putamen), whereas those from the VTA (mesocortical and mesolimbic pathways) terminate on the ventral or limbic striatum (nucleus accumbens).

129

• • • • •

From Latin *back* or *shell*,
In which six layers dwell,
Bulge out and then dive deep,
And all the lobes to keep.

Hint #1:

Contains both gyri and sulci.

Hint #2:

Often called "gray."

Cerebral Cortex

The cerebral cortex makes up the outer gray matter layer of the brain. The cortex itself has six layers, and the proportion of each of these layers differs based on the function and necessity of the area.

Macroscopically, the cerebral cortex is a relatively uniform series of folds and grooves known as gyri and sulci that function to increase the surface area of the cortex. In order, from superficial to deep, the layers of the cerebral cortex are (I) molecular layer, (II) external granular layer, (III) external pyramidal layer, (IV) internal granular layer, (V) internal pyramidal layer, and (VI) multiform layer. Beneath the cell bodies of the cortex begin the myelinated white matter tracts that carry signals to or from the cortex.

The cerebral cortex is a derivative of the telencephalon, which in turn is a derivative of the prosencephalon. There are three main types of cerebral cortex – neocortex, mesocortex, and allocortex. The neocortex is the most phylogenetically advanced type of cerebral cortex with six layers as previously described. This is also referred to as the isocortex and constitutes most of the cerebral cortex.

The mesocortex represents a transition between neocortex and allocortex and contains three to six layers. This type of cortex is present in the limbic parts of the cerebral cortex such as the parahippocampal gyrus, insula, and orbitofrontal lobe.

The allocortex is divided into two subtypes – the archicortex, which is the three-layered cortex of the hippocampal formation, and the paleocortex, which is the three-layered cortex of the olfactory region.

130

• • • • •

In the bottom back of stem,
In floor of four I sit,
Lesion me and you'll feel sick,
Hiccup and/or vomit.

Hint #1:

Contains specialized chemoreceptors.

Hint #2:

Try to "trigger" the right answer.

Area Postrema

The area postrema is located in the posterior inferior portion of the medulla, just inferior to the level of the fourth ventricle. With the absence of a blood–brain barrier (BBB), the area postrema contains specialized ependymal cells that function as chemoreceptors, detecting changes in the blood, especially the concentration of toxins, and reflexively inducing vomiting, earning it the name of chemoreceptor trigger zone.

An inflammatory response against the astrocytes of the area postrema (among other regions) is a typical feature of neuromyelitis optica spectrum disorder (NMOSD). Formerly known as Devic's Disease, NMOSD was once thought to be a manifestation of multiple sclerosis but is now understood as a separate disease entity. Antibodies directed against aquaporin-4 channels cause astrocytic damage, causing injury to the optic nerves, spinal cord, and area postrema. Damage to the area postrema (called "area postrema syndrome") causes intractable episodes of nausea, vomiting, and hiccups. For this reason, patients with area postrema syndrome as the first manifestation of NMOSD often present to gastroenterologists prior to neurologists. It is often not until additional neurologic symptoms manifest that contrasted MRI is performed, showing the typical lesions for NMOSD.

131

· · · · ·

I receive input from frontal lobe,
From me to sixth project,
If lesion me within the pons,
Eyes cannot my side direct.

Hint #1:

Also known as the para-abcucens nucleus.

Hint #2:

Aids in conjugate lateral eye movements.

Paramedian Pontine Reticular Formation

The paramedian pontine reticular formation (PPRF) is a collection of reticular nuclei in the brainstem from the rostral pons to medulla. This collection of nuclei plays an important role in controlling voluntary conjugate eye movements.

The three important nuclei in the PPRF include the nucleus reticularis pontis oralis (NRPO), which contains excitatory burst neurons, the raphe interpositus (RIP), which contains omnipause neurons, and the nucleus reticularis gigantocellularis (NRGC), which contains inhibitory burst neurons.

A descending signal from the frontal eye field to generate contraversive saccades first reaches the superior colliculus ipsilateral to the frontal eye field. The superior colliculus then projects to the contralateral NRPO and activates the excitatory burst neurons. In normal situations, the omnipause neurons in the RIP are tonically active and inhibit the excitatory and inhibitory burst neurons on both sides. The signal from superior colliculus inhibits the omnipause neurons as well.

The excitatory burst neurons in turn activate the abducens nucleus interneurons and motorneurons (contralateral to the frontal eye field). The motorneurons activate the lateral rectus muscle. The interneurons project to the medial rectus subnucleus on the contralateral side (ipsilateral to the frontal eye field) via the medial longitudinal fasciculus. These motorneurons activate the medial rectus muscle of the other eye.

Although selective lesions of the PPRF are unusual because it is a widespread collection of nuclei, these could produce conjugate horizontal gaze palsy ipsilateral to the side of the lesion. This is mostly seen in pontine lesions.

132

• • • • •

Two circles top the midbrain,
Link cortex and cerebel',
If lesion left, right limb will trem',
And movement control not well.

Hint #1:

Also known as nucleus ruber.

Hint #2:

Lesions can cause choreoathetosis.

Red Nucleus

The red nucleus (RN) is a large structure found in the ventral midbrain that contains a magnocellular division and a parvocellular division. It mainly receives afferents from the dentate nucleus of the cerebellum. It has two main outputs – the rubrospinal tract and the rubroolivary tract.

The magnocellular RN is phylogenetically older and is mainly the source of the rubrospinal tract, whereas the parvocellular RN is mainly the source of the rubroolivary tract. The parvocellular RN mostly exists in quadruped animals and is much more advanced in primates (especially humans), whereas the magnocellular RN undergoes regression in primates. The rubrospinal tract only extends up to the cervical spinal cord in humans. In fact, in primates, a complete segregation of the magnocellular and parvocellular portions of RN is seen.

The rubrospinal and corticospinal tracts constitute the two major lateral motor tracts. Unlike the corticospinal tract, the rubrospinal tract decussates in the midbrain tegmentum and terminates in the cervical cord. As mentioned, the magnocellular RN and rubrospinal tract do not have a clearly defined function in humans.

The parvocellular RN is part of the so-called Gullain–Mollaret triangle. This circuit originates at the dentate nucleus (the largest deep cerebellar nucleus). These output cerebellar fibers exit via the superior cerebellar peduncle, which decussates in the midbrain near the red nucleus. Some fibers continue toward the contralateral ventral lateral thalamus and the cortex without synapsing in the red nucleus. Some fibers do synapse in the red nucleus (after decussating in the superior cerebellar peduncle) and then descend via the central tegmental tract toward the inferior olivary nucleus. The olivary nucleus in turn projects back to the cerebellum. These descending fibers via the central tegmental tract

act as feedback to control the olivary output back toward the cerebellum. This dentate-rubro-olivary tract constitutes the Guillain–Mollaret triangle.

Lesions that affect the midbrain tegmentum, including the RN, are associated with ataxia and tremor (Holmes' tremor). A progressive neurodegenerative syndrome affecting the dentatorubral system known as dentatorubro–pallidoluysian atrophy (DRPLA) presents with progressive ataxia, myoclonus, epilepsy, and dementia.

Another intriguing clinical syndrome associated with chronic lesions of the central tegmental tract, known as oculopalatal tremor or myoclonus presents with palatal myoclonus and pendular nystagmus.

133

• • • • •

Ventral to the aqueduct,
Path that helps eye ad'/abduct,
If lesion left, right eye will shake,
And left eye cannot rightward gaze make.

Hint #1:

Connects the sixth and third nuclei.

Hint #2:

White matter tract in the midbrain and dorsal pons.

Medial Longitudinal Fasciculus

The medial longitudinal fasciculus (MLF) is a bilateral paramedian white matter tract located in the midbrain tegmentum and the dorsal pons, which helps to coordinate conjugate eye movements. Its fibers connect the abducens nucleus to the nuclei of the oculomotor and trochlear nerves, allowing for the coordinated adduction and abduction of separate eyes conjugately. It is not only a necessary tract for the functioning of volitional horizontal and vertical gaze but also vital for the vestibulo-ocular reflex.

For conjugate horizontal gaze, signals from abducens nucleus interneurons cross in the midbrain via the MLF to reach the medial rectus subnucleus of the contralateral side. This allows for synchronous activation of the lateral rectus motorneurons on the side of the abducens nucleus and the medial rectus motorneurons on the contralateral side, thus allowing for conjugate horizontal eye movement.

A classic neurological syndrome produced by lesion of the MLF is called internuclear ophthalmoparesis (INO), so named because lesions of the MLF disconnect the abducens nucleus interneurons from the medial rectus subnucleus. This prevents adduction of the eye on the side of the MLF, while the other eye fully abducts. The abducing eye is usually noted to have dissociated abducting nystagmus. This is not a true nystagmus but overshooting saccades, which occur because of Hering's Law of equal innervation. As an equal signal descends into the lateral rectus and medial rectus motorneurons, the medial rectus motorneurons cannot fire fully because of being disconnected. Hence, the signal intensity increases to "force" the weak adducting eye to fully adduct. Because of Hering's Law, an equally strong signal is also sent to the normal abducting eye, causing it to overshoot.

134

• • • • •

From Latin *roof*, I sit atop,
With four humps upon my back,
Behind a tube of CSF,
Both sight and sound we track.

Hint #1:

Two bumps on top, two on bottom.

Hint #2:

Anterior to the superior cerebellum.

Midbrain Tectum (Superior and Inferior Colliculi)

Also known as the quadrigeminal plate, the midbrain tectum, or simply tectum, is the area of the dorsal midbrain sitting posterior to the cerebral aqueduct. On its dorsal surface sit four bumps, the two superior and two inferior colliculi; hence the "quad-" in "quadrigeminal plate." The name tectum is derived from the Latin word for roof, as it serves as the roof for the brainstem. The colliculi themselves are vital junctions in the sensory tracts, with the superior colliculus performing sensory integration of vision and the inferior colliculus performing integration for hearing.

Clinically, the tectum and its immediate surroundings serve as one of the most vulnerable locations in acute neurologic disease. Aside from being a major epicenter for sensation, the tectum also serves as the roof of the cerebral aqueduct, the structure which drains CSF from the third ventricle. Tumors or infarcts within the posterior fossa are well known to compress the tectum against the body of the brainstem, cutting off CSF flow and causing an acute obstructive non-communicating hydrocephalus.

This can cause a classic neuro-ophthalmologic presentation known as Parinaud's syndrome, a triad of impaired upgaze, convergence–retraction nystagmus, and pupillary hyporeflexia (or near-light dissociation). Some academic neurologists consider this to be a tetrad, with eyelid retraction serving as the fourth piece of the puzzle. Patients' impaired upgaze (supranuclear gaze palsy) is a consequence of compression of the rostral interstitial medial longitudinal fasciculus (riMLF), a major conduit of the vertical saccade pathway (and often considered the vertical equivalent to the PPRF). The preferential disability of upgaze is thought to be due to the positioning of the fibers for upgaze within the brainstem, which decussate at the level of the posterior commissure and are thus much more vulnerable to compression.

Convergence–retraction nystagmus (an unfortunate misnomer) is in reality a series of upward-converging saccadic intrusions best elicited by having the patient attempt vertical gaze. The mechanism is proposed to be an injury to supranuclear fibers that inhibit the vergence pathway, causing a release phenomenon.

Pupillary near-light dissociation refers to the fact that, although both light and convergence normally cause pupillary constriction, patients with Parinaud's develop inhibited pupillary light reflex but spared constriction with vergence due to the light reflex pathway, which travels through the pretectum to the Edinger–Westphal nucleus and through the posterior commissure, making it vulnerable to compression.

Finally, patients typically develop eyelid retraction (known as Collier's Sign) due to the compression of the inhibitory fibers that extend to the levator palpebrae superioris as they pass through the posterior commissure.

This is shown in Figure A.1.

135

• • • • •

A connected nuclei system,
Projects to cortex from brainstem,
Fight or flight and wakefulness,
Impaired if brainstem lesion this.

Hint #1:

Located in part in all three segments of the brainstem.

Hint #2:

Helps to regulate a healthy sleep–wake cycle.

Reticular Activating System

The reticular activating system (RAS) is a series of connected nuclei that are located in the brainstem (midbrain, pons, and medulla) and project to many parts of the central nervous system. The nuclei play a very important role in promoting wakefulness and the transition between different states of consciousness.

For wakefulness, a well-organized "dance" of various nuclei secreting various neurotransmitters is required. This involves orexin (or hypocretin) released from the hypothalamus, which promotes the release of acetylcholine from the various cholinergic reticular nuclei at rostral pons and caudal mesencephalon. These cholinergic nuclei include the pedunculopontine nucleus (PPN) and laterodorsal tegmental (LDT) area. In addition, orexin also promotes the release of norepinephrine, histamine, serotonin, and dopamine from various nuclei like the locus coeruleus, tuber cinereum, and raphe nuclei.

Transition from wakefulness to non-REM (NREM) sleep is caused by decreased orexin, acetylcholine, and monoamine levels. This is likely caused by a switch from orexin to GABA in the hypothalamus, produced by the ventrolateral preoptic nucleus.

Lesions of the brainstem, especially at the rostral pons–caudal mesencephalic junction, can produce pathological alteration of consciousness such as coma or stupor because of failure of the normal arousal mechanisms as discussed.

The thalamus could be considered a part of this system, especially the reticular nucleus of the thalamus, which plays an important role in transition from wakefulness to NREM sleep.

136

· · · · ·

Twix thalamus and lens-shaped pair,
And behind the Latin *knee,*
Descending motor fibers come,
To funnel down through me.

Hint #1:

Only two-thirds of me is pure motor.

Hint #2:

"Knee" in Latin is *genu.*

Posterior Limb of the Internal Capsule

The posterior limbs of the internal capsule (PLICs) are bilateral subcortical white matter tracts lateral to the thalamus and medial to the globus pallidus and putamen (together referred to as the lentiform nucleus, derived from the Latin, meaning "lens-shaped").

The internal capsule has an anterior limb, a *genu* (the Latin word for "knee"), and a posterior limb. The anterior two-thirds of the posterior limb of the internal capsule is composed of the descending white matter corticospinal motor tract on its way to funnel down from the cortex through the brainstem. The posterior one-third of the posterior limb of the internal capsule is composed of somatosensory fibers projecting laterally from the ventral posterior thalamic nucleus, called the middle thalamic radiations.

The corticospinal tract in the anterior two-thirds of the PLIC has a somatotopic organization, with the arm fibers being more anterior and medial and the leg fibers being more lateral and posterior. Immediately anterior to the arm fibers is the corticobulbar tract (anterior to posterior – face, arm, trunk, leg).

The anterior limb of the internal capsule is nourished by the recurrent artery of Heubner superiorly and lenticulostriate arteries inferiorly. The genu is nourished by the lenticulostriate arteries. The PLIC is nourished by the anterior choroidal artery superiorly and the lenticulostriate arteries inferiorly.

A classic lacunar stroke syndrome is pure motor hemiparesis, which has a high localizing value to the anterior part of the PLIC. Rarely, it could be present with monoparesis instead of hemiparesis because of selective involvement of arm or leg fibers if the infarct is small enough.

137

• • • • •

Upper medulla, C-shaped bulge,
Twix pyramids and nine,
Where limbs are and intention,
Help make limb movements fine.

Hint #1:

Implicated in multiple systems atrophy.

Hint #2:

Afferents from cortex and spinal cord.

Inferior Olivary Nucleus

The inferior olivary complex is a C-shaped collection of nuclei in the rostral ventral medulla, wedged between the medullary pyramids (medially) and cranial nerves IX, X, and XI (laterally).

The two major cerebellar input fibers are mossy fibers and climbing fibers. The inferior olivary nucleus is the main source of climbing fibers. These fibers enter the cerebellum via the inferior cerebellar peduncle from the contralateral olive. The other source of input via the inferior cerebellar peduncle is the spinocerebellar tracts bringing proprioceptive feedback to the cerebellum.

The olivocerebellar fibers end on the Purkinje cells of the deep cerebellar nuclei. The deep cerebellar nuclei, in turn, send projections outward via the superior cerebellar peduncle to the contralateral ventrolateral thalamus and then to the cerebral cortex. At the same time, some of these projection fibers turn inferiorly near the red nucleus, instead of heading to the thalamus. These fibers descend via the central tegmental tract and project onto the inferior olivary nucleus. This tract acts as the feedback from the deep cerebellar nuclei to the inferior olivary nucleus.

In the early stages of motor learning, the output from the inferior olive to the cerebellum is very active. Over time, as the motor learning is accomplished, the activity of the olivary neurons becomes less.

A chronic lesion of the central tegmental tract in the brainstem can cause deafferentation of the inferior olivary nucleus from the deep cerebellar nuclear feedback. This in turn leads to the development of increased gap junctions between adjacent olivary neurons and synchronized firing. Pathologically, this is referred to as hypertrophic olivary degeneration. Clinically, it manifests as the syndrome of oculopalatal tremor or myoclonus.

A progressive neurodegenerative condition known as multiple system atrophy (MSA) has a cerebellar-predominant phenotype known as MSA-C, also known as olivopontocerebellar atrophy (OPCA), which affects the olivary nucleus and the tracts connecting it to the cerebellum. This is a type of synucleinopathy and is characterized by dysautonomia, progressive ataxia, and Parkinsonism.

138

• • • • •

In lower pons, I sit, a fruit,
With input from each ear,
My fibers send to Latin *hill*,
Localize what you hear.

Hint #1:

Receives input from the cochlear nerve.

Hint #2:

Efferents to the lateral lemniscus.

Superior Olivary Nucleus

The superior olivary nucleus, also called the superior olivary complex, is a small structure located in the caudal ventral pons at the level of the middle cerebellar peduncle that serves as a vital relay station for auditory information. Specifically, the superior olive receives information from the cochlear nuclei via the trapezoid body. The superior olivary nucleus further transmits this information to the inferior colliculus via the lateral lemniscus. This signal is then sent to the medial geniculate nucleus, then to the thalamus, and finally to the auditory cortex.

The superior olive is important for sound localization, using the difference in time sound takes to reach one ear relative to the other to triangulate a sound's origin. It also aids in filtering specific sounds, allowing a listener to focus on the words from a specific speaker while "filtering out" background noise.

An isolated lesion of the superior olivary nucleus is unlikely to produce any clinical hearing difficulty because the central auditory pathway ascends bilaterally and decussates at almost every level after the cochlear nuclei.

139

• • • • •

A group of three on either side,
Subcortically deep white,
Susceptible to high pressure,
Latin *lens* and *striped*.

Hint #1:

Implicated in several movement disorders.

Hint #2:

Largely affected in lacunar infarcts.

Basal Ganglia

The basal ganglia refers to a group of deep subcortical gray matter structures that are connected to widespread areas of the cerebral cortex and play a role in selecting actions.

The neuroembryologic origins of these structures are diverse, as they include telencephalic, diencephalic, and mesencephalic structures. The telencephalic nuclei represent what many refer to as "basal ganglia," which include the caudate nucleus, putamen, and globus pallidus.

The caudate nucleus is medial to the internal capsule, while the putamen and globus pallidus lie lateral. Thus, for anatomic convenience, the putamen, globus pallidus internal, and globus pallidus external are collectively referred to as the "lentiform nuclei," from the Latin word meaning "lens-shaped," given the three of them together approximate the shape of a biconvex lens.

The basal ganglia are phylogenetically much more primitive than the cerebral cortex. In fact, the present-day anatomy and connectivity of various basal ganglia that exists in modern-day humans has been preserved over at least 500 million years of evolution. One of the only extant jawless fish, Lamprey (ancestor of both jawed and jawless vertebrates), has almost the same basal ganglia anatomy, and evolved 500 million years ago during the Cambrian explosion.

The output nuclei of the basal ganglia are the globus pallidus interna (GPi) and substantia nigra pars reticularis (SNpr). These nuclei project to multiple downstream effectors that are central pattern generators (CPGs) for various motor programs (the superior colliculus is one such example for eye movements). The output nuclei are mainly GABAergic and have a high baseline tonic activity, thus inhibiting the downstream effectors from firing and preventing unnecessary movement.

To execute a planned movement, first these output nuclei must be inhibited to disinhibit the downstream effectors. The input nuclei (caudate and putamen) project to the GPi/SNpr. These input nuclei are also GABAergic, but their neurons have a high threshold for firing. Hence, these can only be activated by a strong descending cortical input. Once the input nuclei are activated, they inhibit the output nuclei, thus disinhibiting the downstream CPGs.

This is called the direct pathway, which allows for a selected motor program to be activated. The indirect pathway is the "no-go" pathway and inhibits other motor programs from being activated. This has an additional stop in the globus pallidus externa (GPe), which sends a signal to the subthalamic nucleus (STN), which then projects to the GPi.

As above, it is important to note that the telencephalic nuclei are not the only basal ganglia. The substantia nigra (derived from the mesencephalon) and subthalamic nuclei (derived from the diencephalon) also play vital roles in motor and emotional control and feedback.

This is shown in Figure A.2.

140

• • • • •

I sit behind the olive and
I'm fed from ten and nine,
Send parasympathetics to
The heart, motor to dine.

Hint #1:

Mediates vagal tone.

Hint #2:

Often lesioned in Wallenberg syndrome.

Nucleus Ambiguus

The nucleus ambiguus is the special visceral efferent nucleus for CN IX (glossopharyngeal nerve) and CN X (vagus nerve) that resides in the rostral lateral medulla just dorsal to the inferior olivary nucleus. It supplies motor fibers to the palate and pharynx and mediates the efferent arm of autonomic vagal output, increasing parasympathetic tone in the heart and gut. The name "nucleus ambiguus" refers to the fact that the nucleus is notoriously difficult to locate on dissection or imaging, and its location varies substantially across different species.

The CN IX supplies the stylopharyngeus muscle, while CN X supplies palatopharyngeus, salpingopharyngeus, levator veli palatini, and palatoglossus.

Injury to the nucleus ambiguus is responsible for one of the hexad of clinical features in the infamous Wallenberg syndrome, a classic stroke presentation implying lateral medullary infarction due to occlusion of the posterior inferior cerebellar artery as it branches off the vertebral artery. Damage to the nucleus ambiguus (and therefore bulbar motor fibers) causes dysphagia, dysphonia, hiccups, and a decreased gag reflex.

141

• • • • •

I live in lower brainstem with
Input from arch and tongue,
Help cough and change your breathing rate,
By pH/stretch of lung.

Hint #1:

Essential in the vago-vagal reflex.

Hint #2:

Gets input from 7, 9, and 10.

Nucleus Tractus Solitarius

Located in the medulla, just lateral to the dorsal motor nucleus of the vagus and ventral to the vestibular nucleus, the nucleus solitarius and its tract (together referred to as the "nucleus tractus solitarius" [NTS]) serve as the major sensory input nucleus for the cranial nerves VII, IX, and X.

The NTS is the chief special visceral afferent nucleus for gustatory, respiratory, cardiac, and visceral sensations. This nucleus has rostral, intermediate, and caudal divisions.

The chorda tympani, a tributary to CN VII, carries taste innervation from the anterior two-thirds of the tongue, while the glossopharyngeal nerve (CN IX) transmits taste from the posterior one-third. This sensation is mostly received in the rostral NTS.

Visceral sensory information, via gastrointestinal vagal afferents, terminate in the intermediate NTS. Cardiorespiratory afferents via the CN IX and X terminate in the caudal NTS.

The baroreceptor reflex is mediated by CN IX and X afferents from the carotid sinus and aortic arch. These receptors are activated by increased arterial pressure. The incoming signal is relayed to the NTS, which in turn projects to inhibitory interneurons in the caudal ventrolateral medulla. This area normally has sympathoexcitatory neurons in the rostral ventrolateral medulla. Inhibition of these neurons decreases peripheral arterial resistance, thus decreasing blood pressure. In addition, the NTS also projects to the cardiac preganglionic parasympathetic neurons in the nucleus ambiguus. These neurons project to the cardiac SA node via CN X and induce bradycardia, also decreasing blood pressure.

For respiratory regulation, afferent signals from carotid bodies triggered by hypoxemia or hypercapnia reach the NTS, which in turn projects to the inspiratory group of neurons in the pre-Bötzinger complex.

Finally, regulation of gastrointestinal motility occurs through autonomic visceral sensation feedback loops moving from the gut to the brain or spinal column, and back to the gut. The best-studied reflex from the nucleus solitarius is the vago-vagal reflex, where distention of the stomach feeds information up to the nucleus solitarius, which in turn signals the dorsal motor nucleus of X to increase gastric motility and acid secretion.

142

• • • • •

Inside a lateral placed groove,
Side lobe atop I sit,
With input from Latin *like-knee*,
To hear and interpret.

Hint #1:

Last stop before the language center.

Hint #2:

Think temporal lobes.

Primary Auditory Cortex

The primary auditory cortex is located within Heschl's gyri, which are transverse temporal gyri located on the posterior aspect of the superior surface of the superior temporal gyrus within the sylvian fissure. This corresponds to Brodmann area 41. This region receives input from the medial geniculate nucleus (*genu* in Latin meaning "knee"). Planum temporale refers to the part posterior to the transverse gyri up to the termination of the sylvian fissure.

The auditory association cortex is just lateral to the primary auditory cortex, located on the superior and lateral surface of the superior temporal gyrus. This refers to Brodmann areas 42 and 22.

Epileptic seizures affecting the auditory cortex present with auditory auras, characterized by simple sounds such as buzzing, ringing, humming, and so on. The aura may or may not be lateralized to one side. The pitch of the auditory aura also implies where in Heschl's gyrus the seizure is coming from; high pitch localizes to the anterior segment of the auditory cortex, while low pitch localizes to the posterior segment. Despite providing excellent help with localization, an auditory aura has poor lateralizing value because around half of auditory fibers decussate to the contralateral auditory cortex. So, a buzzing aura in the right ear may localize either to the left or right primary auditory cortex.

Electrical cortical stimulation studies have also confirmed these findings. Stimulation of area 41 is more likely to induce broadband noise-like hallucinations. Stimulation of area 42, area 22, the planum temporale, and the superior temporal gyrus are more likely to induce auditory illusions.

143

• • • • •

From Latin *belt*, just off midline,
Near falx I am located,
Emotional response to pain,
Most common herniated.

Hint #1:

Part of Papez Circuit.

Hint #2:

Perfused by the anterior cerebral artery.

Cingulate Gyrus

Residing in the depths of the interhemispheric fissure, the cingulate gyrus gets its name from the Latin word for "belt," *cingulum*, as it wraps around the corpus callosum. The cingulate gyrus is separated from the superiorly located superior frontal gyrus by the cingulate sulcus and from the inferiorly located corpus callosum by the callosal sulcus. The cingulate sulcus has a marginal extension that separates the superior frontal gyrus and cingulate gyrus from the posteriorly located precuneus. The arterial supply of the cingulate gyrus is derived from the pericallosal arteries, which are branches of the anterior cerebral artery.

The cingulate gyrus (aka cingulate cortex) is broadly divided into the anterior cingulate cortex (ACC), middle cingulate cortex (MCC), and posterior cingulate cortex (PCC). Two imaginary vertical lines, through the anterior commissure (VCA) and the posterior commissure (VCP), roughly separate the cingulate cortex into these three divisions.

The ACC is further subdivided into a subgenual and a pregenual sector (sACC and pACC). The MCC is divided into an anterior and posterior division as well, and each one of those is further divided into a ventral and dorsal subdivision.

The ACC is mostly known to play a role in affective (emotional) responses and autonomic responses. Electrical stimulation studies of sACC (Brodmann area 25) have been known to induce hypotension.

Electrical stimulation studies in epilepsy patients have demonstrated a variety of complex motor behaviors elicited by stimulation of the MCC. In fact, the MCC is thought to have an "actotopic" organization of complex motor behaviors, which become increasingly more complex in a ventral to dorsal gradient. Meanwhile, the pMCC and PCC are mostly known to have vestibular and visual functions.

Of particular importance to memory function is the cingulum, a white matter tract that serves as part of the Papez Circuit by transmitting information passed by the anterior nucleus of the thalamus to the cingulate cortex and then to the parahippocampal gyrus.

Given its near midline location, the cingulate gyrus is often the first structure displaced during focal cerebral edema, such as in the case of middle cerebral artery territory infarction. In the case of edema in the structures lateral to the cingulate gyrus, it becomes pressed against the falx cerebri and then underneath it, presenting the classic "midline shift" typical of subfalcine herniation. Depending on the degree of shift, the cingulate gyrus may also compress the branches of the anterior cerebral artery, leading to secondary infarction.

This is shown in Figure A.1.

144

• • • • •

Two gyri of two different lobes,
Blood from anterior,
Injure left, leg weakness, and
Numbness, on right, occur.

Hint #1:

Positioned just off midline.

Hint #2:

At the end of the cingulate sulcus.

Paracentral Lobule

Located within the interhemispheric fissure, the paracentral lobule straddles the frontal and parietal lobe as an extension of both the precentral and postcentral gyri. It is here that both motor function (precentral) and sensation (postcentral) localize for the contralateral lower extremity. This area is perfused by divisions of the anterior cerebral artery (ACA), typically the callosomarginal artery.

On the medial surface of the brain, the superomesial extension of the central sulcus is situated immediately anterior to the marginal extension of the cingulate sulcus. The postcentral gyrus exists between the central sulcus and the marginal sulcus, and it represents the primary somatosensory cortical representation of the leg, foot, and genitals.

Immediately anterior to the central sulcus is the precentral gyrus, which constitutes the primary motor representation of the leg and foot. Inferior to this area and continuing anterior to it is the supplementary sensorimotor area (SSMA), which has its own homunculus with the legs posterior (facing the leg of the primary motor) and the arm and head anterior.

Ischemia and infarction of the paracentral lobule is a hallmark of ACA occlusion, which typically presents with hyperacute onset of contralateral weakness and hemisensory loss in the lower extremity out of proportion to the face and upper extremity. Infarction of the paracentral lobule can also produce urinary incontinence, though the mechanism is not completely understood. Additionally, patients with ACA-territory infarcts often demonstrate clinical signs of frontal lobe damage including abulia (a triad of low spontaneous verbal output, prolonged speech latency, and paucity of response).

This is shown in Figure A.1.

145

• • • • •

A vertical fluid-filled tube,
That stretches fourth to cone,
Lined by one-layer columned cells,
I am dilation prone.

Hint #1:

Think hydromyelia.

Hint #2:

If dilated causes cape-like sensory loss.

Central Canal

Extending from the fourth ventricle down to the conus medullaris, the central canal of the spinal cord is a very narrow CSF space bordered on all sides by a single layer of columnar ependymal cells. Recall that the human central nervous system first develops in utero in a process known as neurulation, and during the fourth week of development, the caudal aspects of the neural plate and neural tube fold in on themselves and zipper shut, forming a neural canal which will eventually become the central canal. The neural tube, a single layer of columnar neuroepithelial cells, then differentiates into all neurons, oligodendroglia, and astroglia of the spinal cord before differentiating into that one layer of columnar cells that lines the central canal.

When CSF flow through the central canal becomes obstructed (either by bodies within the central canal or by extrinsic compression of the cord), backup of CSF causes dilation of the central canal known as hydromyelia. Recall that the spinothalamic tract, which carries sensory fibers for pain and temperature, decussates across the spinal cord in the ventral white commissure, adjacent to the central canal. Thus, patients with hydromyelia classically first develop loss of pain and temperature sensation in a "cape-like distribution" as the syrinx slowly expands downward.

146

· · · · ·

A bump on back of thalamus,
Where optic tract there ends,
Synapse in me, and second nerve,
Gets what retina sends.

Hint #1:

Perfused from both anterior and posterior circulation.

Hint #2:

If lesioned produces a unique anopia.

Lateral Geniculate Nucleus

Lying at the posterior edge of the thalamus, the lateral geniculate nucleus (LGN) is an integral part of the human visual pathway. The LGN is the first relay station for the optic radiations, which synapse here and give information to the second-order neurons that project to the primary visual cortex.

The LGN is a six-layered structure; each layer either receives input from the ipsilateral eye or the contralateral eye. Remember, the visual information in each optic nerve splits at the optic chiasm, where the temporal fibers continue undecussated, but the nasal fibers decussate at the chiasm. Hence, each optic radiation has fibers from each eye.

In addition, the two most ventral layers of the LGN are referred to as magnocellular layers. These primarily receive afferents from the parasol retinal ganglion cells. The remaining four layers are referred to as the parvocellular layers. These receive information from the midget retinal ganglion cells.

A lesion of the LGN can cause contralateral homonymous hemianopia, and incomplete lesions cause incomplete visual loss. The LGN receives dual arterial supply; if viewed in the coronal plane, the middle part of the LGN is nourished by the posterior choroidal artery (a branch of the posterior cerebral artery), whereas the lateral and medial horns are nourished by the anterior choroidal artery (a branch of the internal carotid artery).

Anterior choroidal artery infarction presents with quadruple sectoranopia, which refers to two congruent upper and lower wedges of each hemifield. Posterior choroidal artery infarction presents with horizontal sectoranopia, which refers to two horizontal congruent wedges of each hemifield.

147

• • • • •

A bi-direction synapse point,
Thalamically placed rear,
If lesioned cannot localize,
Sounds once they come in ear.

Hint #1:

Between the inferior colliculus and auditory cortex in track.

Hint #2:

Helps you react quickly to potentially dangerous noises.

Medial Geniculate Nucleus

The medial geniculate nucleus (MGN), found in the posterior thalamus, serves as the relay point for auditory information between the inferior colliculus and the auditory cortex. Additional projections from the MGN diverge to the amygdala and mediate classically conditioned responses to auditory stimuli.

The MGN has three main subdivisions – ventral (MGV), dorsal (MGD), and medial (MGM). The MGV is the core subdivision and projects mainly to the primary auditory cortex. Unlike the MGV, the auditory responses in the MGD can be modulated by non-auditory inputs. The MGD and MGM receive afferents from multiple other sources in addition to the inferior colliculus and also project to other targets besides the primary auditory cortex, such as the amygdala. The MGV receives afferents mainly from the inferior colliculus and projects to the primary auditory cortex.

The MGN has also been implicated in sound localization like the superior olive. Animal studies have shown that 20–30% of cells in MGN are only responsive to stimulation in the contralateral ear.

Isolated lesions of the MGN in humans are rare and are not necessarily associated with any specific symptoms although sound localization might be affected. Rarely, auditory illusions have been reported.

148

• • • • •

A crossing-point of white matter,
Afferent signals send,
Their second-order nerves must pass,
Before cord they ascend.

Hint #1:

Runs anterior to the central canal.

Hint #2:

Think spinothalamic tract.

Ventral White Commissure

Unlike the motor fibers of the corticospinal tract and the vibratory/proprioceptive fibers of the dorsal columns, the small pain/temperature fibers of the spinothalamic tract (STT) do not decussate within the medulla, but rather within the body of the spinal cord itself. Fibers carrying pain and temperature sensation are small and unmyelinated. These transmit signals via the pseudounipolar neurons in the dorsal root ganglia into the dorsal horn of the spinal cord. Most of the fibers synapse in the dorsal gray matter of the spinal cord (substantia gelatinosa). Some fibers do not synapse immediately but instead ascend or descend two to three spinal segments via Lissauer's tract before synapsing.

After synapsing, the second-order axons cross the ventral white commissure and continue up the contralateral spinal cord and brainstem as the STT, until they reach the ventral posterolateral nucleus of the thalamus. These axons synapse here, and then the third-order neurons project to the cerebral cortex.

This unusual decussation within the spinal cord via the ventral white commissure renders these fibers susceptible to injury with ventral intrinsic cord lesions. Because of its proximity to the central canal, a small dilation of that canal can impinge on the second-order axons in the ventral white commissure, as is seen with syringomyelic syndrome.

149

• • • • •

Ascending second-order group,
Of nerves post-decussate,
From medulla to thalamus,
Help sense move and vibrate.

Hint #1:

Part of the dorsal columns tract.

Hint #2:

Think medial medulla.

Medial Lemniscus

The medial lemniscus is a white matter tract in the brainstem that is made up of second-order somatosensory axons carrying touch, vibration, and proprioception.

Sensory fibers carrying light touch, proprioception, and vibration are usually large and heavily myelinated. The first-order neurons are the pseudounipolar neurons in the dorsal root ganglia. These fibers enter the dorsal horn of the spinal cord but, unlike the pain and temperature fibers, do not synapse with the dorsal root gray matter. Instead, these fibers simply ascend in the dorsal funiculus of the spinal cord as the medial funiculus gracilis (lower extremity) and the more lateral funiculus cuneatus (upper extremity).

These first-order axons synapse with the second-order neurons in the nucleus gracilis and nucleus cuneatus in the caudal medulla. The second-order axons immediately decussate ventrally via internal arcuate fibers and then ascend more ventral and medial as the medial lemniscus.

The medial lemniscus then synapses with the third-order neurons in the ventral posterior lateral thalamus before projecting to the cerebral cortex.

Lesions affecting medial brainstem often present with contralateral vibratory/proprioceptive sensory loss because of the involvement of medial lemniscus.

150

● ● ● ● ●

With roots from sacral two through four,
Latin *to be ashamed*,
Motor and sensory to floor,
A mess if nerve is maimed.

Hint #1:

Helps with urinary/fecal continence.

Hint #2:

Important for sexual function.

Pudendal Nerve

Derived from the Latin word for "parts to be ashamed of," the pudendal nerve provides sensorimotor innervation to the penis, clitoris, scrotum, labia majora, inferior vaginal canal, and anus. It plays a vital role in gastrointestinal and genitourinary function.

Emerging from the S2–S4 rami of the sacral plexus, the pudendal nerve travels next to the lateral wall of the pelvic cavity in a neurovascular bundle with the internal pudendal artery and vein. It courses out of the pelvis by passing under the piriformis muscle through the greater sciatic foramen before wrapping around the sacrospinous ligament next to the ischial spine. It then dives down through the lesser sciatic foramen and hugs the obturator internus muscle in the pudendal canal (formerly called Alock's canal). Here, it gives off its three major branches – the inferior rectal nerve, the perineal nerve, and the dorsal genital nerve (also known as the dorsal nerve of the penis/clitoris).

Through the function of the inferior rectal nerve, the pudendal nerve provides control of the external anal sphincter and sensation to the anal canal below the dentate line. From the perineal nerve, the pudendal nerve provides motor control of the external urethral sphincter, regulates clitoral/penile tone by restricting venous output, coordinates the perineal muscles to perform penile ejaculation, and carries sensation from most of the urogenital triangle. Finally, through the pure sensory dorsal genital nerve, the pudendal nerve transmits sensation from the body and glans of the penis or clitoris, promoting sexual function by serving as the afferent arm of the erectile pathway.

Clinically, the pudendal nerve may suffer either acute or chronic insult. Acute pudendal neuropathy is an uncommon complication of vaginal delivery typically due to stretch injury, which results in loss of bowel and/or bladder continence and may mimic symptoms of a spinal epidural hematoma acquired after epidural anesthesia. Chronic pudendal nerve entrapment injury, known as pudendal neuralgia, is a similarly rare disorder typically caused by prolonged sitting or repetitive hip flexion. Patients typically suffer dull perineal pain that progressively intensifies throughout the day and is worse with sitting down, often appearing clinically like chronic prostatitis.

Appendix

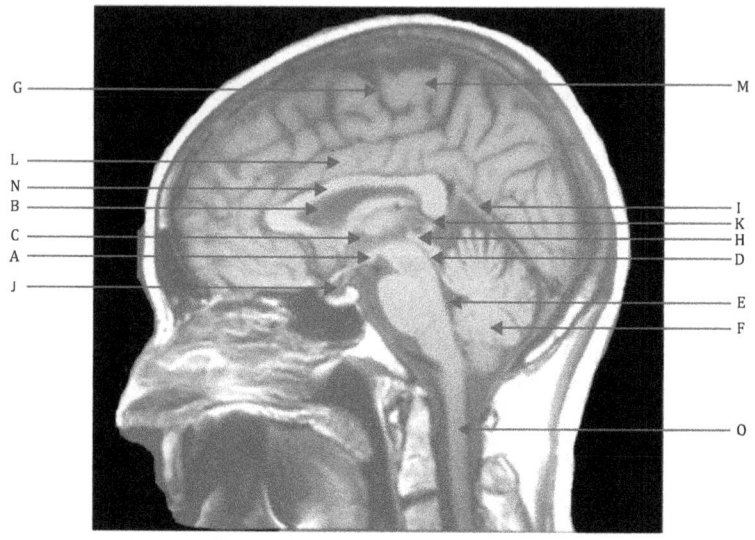

A Mammillary Body
B Lateral Ventricles
C Third Ventricle
D Cerebral Aqueduct
E Fourth Ventricle
F Cerebellum
G Central Sulcus
H Midbrain Tectum (Superior and Inferior Colliculi)
I Tentorium Cerebelli
J Pituitary Gland
K Pineal Gland
L Cingulate Gyrus
M Paracentral Lobule
N Corpus Collosum
O Spinal Cord

Figure A.1 MRI brain (midsagittal view, T1 sequence)

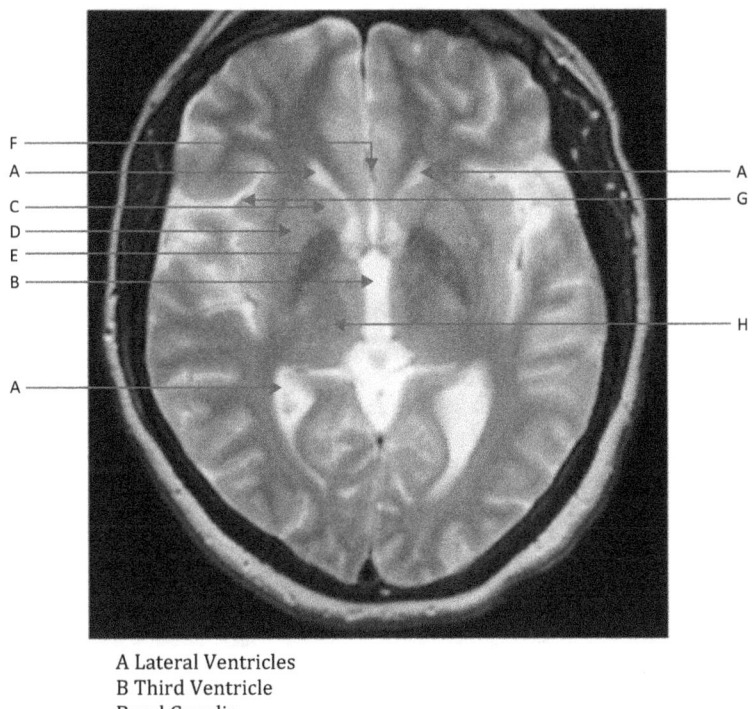

A Lateral Ventricles
B Third Ventricle
Basal Ganglia:
  C Caudate (head)
  D Putamen
  E Globus Pallidus
F Anterior Cerebral Artery
G Middle Cerebral Artery
H Thalamus

Figure A.2 MRI brain (axial view through the basal ganglia, T2 sequence)

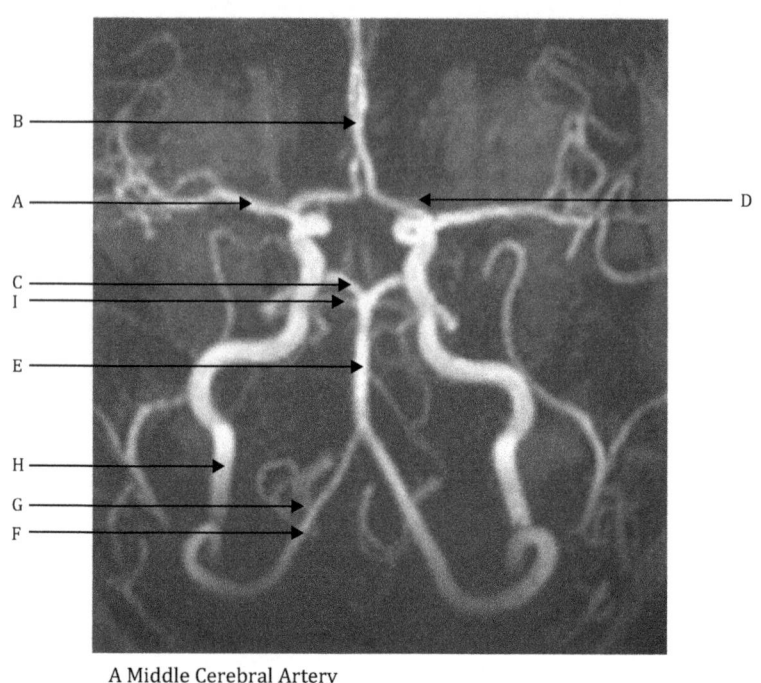

A Middle Cerebral Artery
B Anterior Cerebral Artery
C Posterior Cerebral Artery
D Anterior Choroidal Artery
E Basilar Artery
F Vertebral Artery
G Posterior Inferior
  Cerebellar Artery
H Internal Carotid Artery
 I Superior Cerebellar Artery

Figure A.3 Intracranial MR angiogram (anterior–posterior projection)

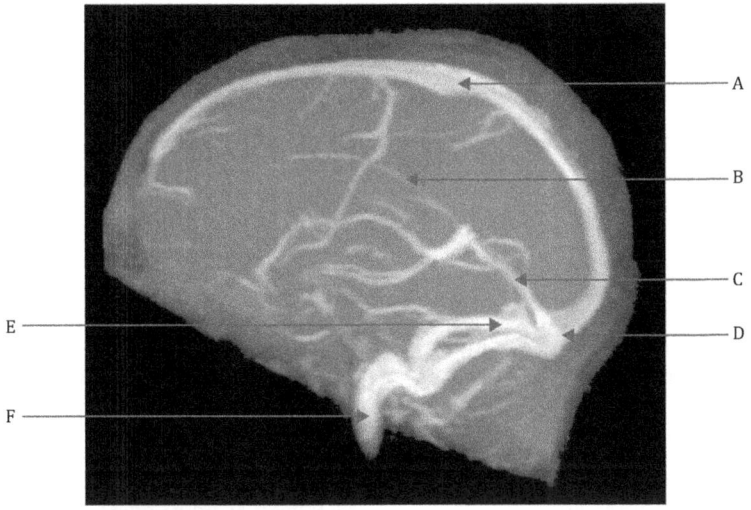

A Superior Sagittal Sinus
B Inferior Sinus
C Straight Sinus
D Torcula (Confluence of Sinuses)
E Transverse Sinuses
F Internal Jugular Vein

Figure A.4 Intracranial MR venogram (lateral view)

Images in the appendix are provided by *Neuroimaging in Neurology: An Interactive Approach,* with permission from Dr. David Preston.

Index

348

For EU product safety concerns, contact us at Calle de José Abascal, 56–1°,
28003 Madrid, Spain or eugpsr@cambridge.org.